AMERICAN SAILORS & MARINES DURING THE REVOLUTIONARY WAR

IN THEIR OWN WORDS

By

Edward Leo Semler Jr.

First Edition: 2023

Library of Congress Control Number: 2023914159

ISBN: 978-1-7376472-3-2

Printed in the United States of America

City of Publication: Schulenburg, Texas

Cover layout by Edward Leo Semler Jr.

To all these veterans and their family's

Table Of Contents

Introduction

As I researched writing several other books on the Revolutionary War, I was intrigued by the naval pension claims I read. They really didn't fit into what I was writing at the time, and I felt their stories needed to be told.

When I started to compile my list of naval pension claims for this book, I also found it interesting that when I came across a vessel mentioned in a claim, there wasn't a lot of information available on that ship when conducting an internet search. So, I thought by compiling these stories, it would also give the reader an insight as to what these ships were like.

But as I came across more and more vessels listed in claims, I soon realized that I wouldn't be able to include them all in this book. Not only is there over 60, sort of "official," ships listed by various sources as serving in the Continental Navy, but in my research, I came across many more vessels that fought for America's Independence. Many of these were state/colony owned, privateers, seized war prizes, armed merchants, and armed vessels that were not officially listed as being pressed into service.

I think you will find that the sampling I have provided in this book gives a great insight into Americas first naval vessels. Because, even though the following chapters are about Continental Navy ships, the men in these chapters usually describe their service on multiple vessels, some Continental Navy and others in the categories previously mentioned.

The founding of the United States Naval Service & Continental Marines began with their formation in 1775 as the result of the war for independence from Britain. Before this, colonies had stood up navies under their own control. But in October of 1775 the Continental Congress passed a resolution to stand up a Colonial, or United States, navy and marines. Initially this consisted of building fighting ships. But as I mentioned earlier, this list of ships under Continental control grew to include state/colony ships, privateers, converted merchant, and seized war prize vessels. At the end of the Revolutionary War the Continental Nay and Marines were disbanded until the 1790's when they were both reinstated.

Federal pension claims, or applications, came in several stages during and after the war. In 1780 there was an initial pension act established for disabled veterans and widows of fallen soldiers. Later in 1818, 1820, and 1822 it was opened to those who fought more than 9 months in the Continental Army, Navy, or Marines. And finally, there was the act in 1832 that opened pensions to everyone else who served more than 6 months, to include the militias and those serving as Indian scouts and spies. But unfortunately, most pension applications submitted before 1800 were lost in a fire at the War Department.

From my research I found that most existing claims were filed under the 1832 pension act. And these typically offer the best information about a veteran's service. They required a lot more proof of service along with the names of the officers the claimant served under. So, applicants were very detailed with describing their service, so their claim would be approved.

But unfortunately, this meant that it wasn't until some 50 years after the Revolutionary War that a lot of veterans were allowed to apply for a pension. Now in their 70's, and beyond, they had to try and remember what happened almost 50 years earlier when they fought for their country. Sadly, a majority of these claims just mentioned the basics facts of their service, such as their unit, people they served under, and the years. Some are not in first person and the claimant is the veterans' widow or other family member having the right to file a claim. But every now and then I came across a claim that had an interesting first-person account about their service. And some were just unbelievably detailed.

If you think about it, there really had to be a lot of factors that came together for these stories to have survived. First, the veteran had to live into old age to make it to 1818 or 1832 and be able to have their story documented. Second, the claim had to be written legibly by the agent documenting the claim. And third, the veteran had to have the personality to relay more than just the basic details needed to prove the claim. I guess it could also have fallen on the agent to have pulled the information out of the veteran. But I think you'll see in the stories I am about to re-tell that these men had personality, and great memories. And as I relay their stories, I am going to be quoting them just as the veteran presented their story to the claim agent and the agent wrote it down. I'll preface the story with any details that may need explaining so as you read their story, you'll have an idea of what they are talking about. And I'll add "author notes" to help explain anything I feel the reader may be unfamiliar with. But I will not try and correct any historical inaccuracies in their stories. For whom am I to correct someone who was actually there.

You may find these claims a bit hard to read at first. They had a slightly different vocabulary when these were written and they tend to be repetitive and run on. But once you start to read them, you'll find them easy to understand. Just image how I felt having to read the quill & ink cursive that was used to write these. Defiantly not John Hancock penmanship!

I have included their entire pension declaration when possible. To include the beginning statement that was required for the 1832 claims. This opening has important information on the individual, such as place of residence and age. In the case where the individual has lengthy land service before his sea service, I have omitted the land service. And in these cases, I notify the reader that I am doing so. Also, when a sailor or marine has served on multiple vessels in one claim, I place their story within the vessel chapter that appears to have the majority of their service.

This is also by no means a complete list of the crews that served on these ships. I am just mentioning the sailors and marines who had pension claims that expanded on their service while aboard these vessels. The claims that I came across that only briefly mention their service or vessel, I have omitted.

And finally, I have enclosed two diagrams on the following pages to help the reader. The first showing the location of sails, which is mentioned frequently in this book. Although sloop, schooner, brigantine, and frigate sail plans differ, this will give you a general idea of what these men are talking about. And second, a profile view of some of the ships referred to by these men. Where possible I have also included an illustration of the vessel at the beginning of the ships chapter.

So, lets shove off, and as these men would say, "go on a cruise!"

Sails of a Ship

1. Flying jib
2. Outer jib
3. Inner jib
4. Fore topmast staysail
5. Fore royal
6. Fore upper topgallant sail
7. Fore lower topgallant sail
8. Fore upper topsail
9. Fore lower topsail
10. Fore sail, or fore course
11. Main royal staysail
12. Main topgallant staysail
13. Main topmast staysail
14. Main skysail
15. Main royal
16. Main upper topgallant sail
17. Main lower topgallant sail
18. Main upper topsail
19. Main lower topsail
20. Main sail, or main course
21. Mizzen topgallant staysail
22. Mizzen topmast staysail
23. Mizzen royal
24. Mizzen topgallant sail
25. Mizzen upper topsail
26. Mizzen lower topsail
27. Crossjack (hanging in gear—not set)
28. Spanker

Ship
Bark
Full-rigged Brig
Hermaphrodite Brig
Top-sail Schooner
Fore & aft Schooner
Sloop

USS Alliance

The *Alliance* was a 36-gun Frigate purchased by the Continental Congress in 1777. She was manned with a crew of around 100 men and sailed under an American flag until she was sold in 1785.

Her captains were said to be John Paul Jones, Pierre Landais & John Berry. I say "were said to be" in this and subsequent chapters because I found that the historical record and pension claims differ on who actually commanded these vessels.

Alliance - Directory of Naval Fighting Ships.

Joseph Frederick

Commonwealth of Massachusetts

I Joseph Frederick aged seventy years, born in Portugal but for fifty four years last past a resident citizen of the United States, and now resident in Starks in the District of Maine testify and declare upon oath that in December 1777 or January 1778 I entered as a Boatswains Mate on board the ship Alliance, Capt. Peter Lander commander in the naval service of the United States, and I served on board in said capacity one year and eight months, was in company the ship Bonhomme Richard, Capt. Paul Jones commander in the engagement and aforesaid in capturing the ship Serapis and ship Scarborough, two English ships; and I received my final verbal but honorable discharge from Capt. Lander at Boston according to the best of my recollection in August 1779.

Author's note – In declarations by his wife and a friend, they state that Joseph Frederick was wounded in the leg during the engagement with the *Serapis.*

Joseph Frederick's pension application was approved and he received $8.00 per month for service under the 1818 Pension Act.

He passed away on the 7th of June 1821.

George Raymond

State of New York, Kings County

On this twenty fifth day of October in the year one thousand eight hundred and thirty AD personally appeared in open court before the Municipal Court in the village of Brooklyn in said county and state now sitting George Raymond, a resident of Brooklyn aforesaid aged seventy three years & who being first duly sworn according to law doth on his oath make the following declaration in order to obtain the benefit of the pension made by the Act of Congress passed June 7 1832.

That he was born in Norwalk, Fairfield County, State of Connecticut on the First day of January in the year 1759 – That he enrolled in the militia and served under Captain Seth Seymour - In the same year was drafted and served a campaign at Fort Independence in the State of New York – That he entered in the same year on board the privateer Independence commanded by James McGee as Captains Clerk and served on board said privateer four months. – That while on board said privateer two of the enemy's vessels were captured and he came into New Bedford in one of them as Prize Master. – That in the spring of 1779 he entered on board the privateer Brig General Washington commanded by Captain William Rogers, that while on said vessel they captured the English vessel Jonathan, Captain Townsend, in which he came into Boston as Prize Master – That in the same year he entered and served four months on board the vessel Mars, a privateer in the capacity of Captain of marines. – That he afterward entered on board the sloop Independence commanded by Captain Francis Brown as first officer and was captured and carried into England. That he was exchanged and getting to Holland at Texel about the last of November 1779.

He went aboard the U.S Frigate Alliance as a volunteer in the capacity of midshipman, said Frigate commanded by John Paul Jones Esq. That from said station they proceeded to Coruna in Spain, thence to L'Orient in France, all this was under the command of said John Paul Jones, which was about six months. After which Captain Landais, the former Capt. of the ship was re-instated to his command. That he continued on board said ship under Captain Landais until about the middle of September 1780 when he was discharged in the pay of Sailing Master Mate at the Port of Boston. The time he was under the command of Captain Landais the 1st Lieutenant was James Diggs, the 3rd Lieutenant Sailing Master Buckley, Captain of Marines Parks and Lieutenant of Marines Warner. That at a Court Inquiry upon Captain Landais he with Arthur Lee Esq were witnesses against him.

Author's note – Captain Landais was brought up on charges, which you will read more about in John Kirby's pension application, which is in the *USS Bonhomme Richard* chapter.

That in 1781 he sailed in the American Brig Baltimore and was captured by Admiral Rodney at the Eustacea and carried to

England, was exchanged and taken prisoner again in 1782 and continued until peace took place in 1783.

Author's note – George Raymond's pension application was approved and he received $85.50 per year for service under the 1832 Pension Act.

He passed away on the 4th of May 1844.

USS Andrew Doria

The *Andrew Doria* documents a major milestone in American naval history. She was presented with the first salute to an American naval vessel from a foreign country. The historical salute was given by the Caribbean, or West Indies, country of St. Eustatius on the 16th of November 1776. As *Andrew Doria* entered the foreign port, she fired her guns in salute, and received the same from the fort protecting the harbor.[1]

She was a converted merchant ship purchased in 1775, had 14 guns, and a crew of around 110 men. In 1777 she was scuttled at Red Bank, New Jersey to prevent being captured by the British.

Her captains were said to be Nicholas Biddle & Isaiah Robinson.

Andrew Doria - Alchetron.com

James Josiah

James Josiah a citizen of the United States, resident in the County of Philadelphia, State of Pennsylvania, aged sixty-eight years, on his oath declare: That in or about the month of October One Thousand Seven Hundred and Seventy Five he was appointed first lieutenant of the Brig Andrew Doria fitted out at Philadelphia and commanded by Captain Nicholas Biddle. That he sailed with the fleet under Commodore Hopkins on the expedition against New Providence and after the surrender of that place returned in the said Brig with the fleet to New London.

Author's note - New Providence is an island in the Bahamas.

That he sailed in the same vessel as first lieutenant on a cruise upon the banks of Newfoundland, and the Andrew Doria having captured two transport ships with Scotch troops on board, the command of one of them was given to the deponent with orders to proceed to Newport or New London, But off Nantucket he was taken by the British Frigate Cerberus and detained in her three months, was afterwards put on board the prison ship Whitby at New York, where he remained near five months when he was exchanged on the 24th December 1776.

Author's note - The *Andrew Doria* ran into the British transport vessels *Oxford* and *Crawford* loaded with Scottish troops. James Josiah was placed aboard *Crawford*.[1]

The ill treatment received by this deponent while in the hands of the enemy was made the subject of complaint by Captain

Biddle as well to the British admiral at New York and to the American government at the council of congress of 7 August 1776, there the notice taken of the subject by that body. On the 20th August 1776 while a prisoner the deponent was promoted to the rank of captain in the navy of the United States and appointed to the command of the ship Champion lying in Philadelphia, and was engaged in her at the siege of Fort Mifflin and Red Bank.

Author's note - Champion was armed with 8 guns.

After the evacuation of the fort, the said ship with others was burnt by order of a council of war and the deponent proceeded with his crew by boat to Bordentown where he fell out of the service and entered the private service.

Author's note - As the British closed in to capture Philadelphia, those vessels that were unable to break out were burnt to prevent capture.

After the evacuation of Philadelphia by the British in June 1778, the deponent obtained a furlough to command a privateer and made several voyages in the private service, having orders from the navy board to report himself on his return to port, which he conformed to do until the Peace, never having been discharged from the service. When the deponent entered the private service he gave up his commission applicable to the general order of the navy board and did not again receive it having continued in said service until the conclusion of the war, though able and ready to do duty when called upon by government.

Author's note – After the burning of their vessels several of the captains were given furloughs to enter private service as long as they checked in with the navy whenever they came into port to see if they were needed for active duty service.

James Josiah's pension application was approved and he received $20.00 per month for service as a captain under the 1818 Pension Act.

He passed away on the 18th of September 1820.

His pension claim deposition was given on the 14th of January 1820, so he only was able to draw about 8 months of that pension. But his wife Elizabeth, who was 79 years old at the time, filed a claim for widows later on in 1843 and began to draw $552.00 per year on his pension.

Isaac Dewees

United States

Delaware District

Deposition of witnesses produce & sworn or affirmed and examined before me John Higher District Judge of the United States for Delaware District on the claim of Isaac Dewees, late a corporal of Marines in the Navy of the Revolutionary War and now a resident of the district aforesaid, for a pension in performance of the provisions of an Act of Congress, entitled "An Act to provide for certain persons engaged in the land and naval service of the United States in the Revolutionary War;" approved on the 18th day of March 1818.

The aforesaid Isaac Dewees aged sixty seven years and upwards, personally appeared before me and being affirmed in due form of law (he being conscientiously scrupulous of taking oath) declares as follows, That in the month of November in the year of our Lord one thousand seven hundred and seventy-five this declarant enlisted in the naval service of the United Colonies then so called, at the city of Philadelphia as a marine and was put on board a brig of war, then belonging to the said United Colonies called the "Andrew Doria" under the command of Capt. Nicholas Biddle; that the first expedition or cruise the said brig sailed on was as one of the fleet of Commodore Hopkins to New Providence, one of the Bahama Islands, and belonging to the crown of Great Britian; that on the arrival aforesaid fleet at New Providence, both forts surrendered to said fleet and gave up all their munitions of war, which were brought home to said Colonies and landed safely at New London in the state of Connecticut; that after aforesaid expedition to New Providence, the brig aforesaid continued to cruise on the American coast and captured a number of British prizes, through the summer of 1776 until the month of August 1776, at which time Capt. Biddle left command aforesaid brig and he came under the command of Capt. Robinson, under whose command he made a voyage to St. Eustatia for naval and military stores for account of the then Continental Congress; that on the voyage from St. Eustatia to the United States said brig captured five prizes, the one a sloop named the "Race-Horse" the other a sloop named the "Tom Jones;" that the applicant was put on board the Tom Jones under Lieutenant Joshua Barry (since Commodore) who was appointed by Captain Robinson her prize master; that said prize got safe into Chincoteague Inlet in the Commonwealth of Virginia, at which place this applicant was discharged from said prize, the term of his enlistment, it being one year having expired; that the discharge was in the month of January in the year 1777, and was a verbal discharge given to this applicant by one Thomas Cummins, he being Lieutenant Barry's mate of said prize, the Lieutenant himself having gone to Baltimore.

Author's note - Isaac Dewees' pension application was approved and he received $8.00 per month under the 1818 Pension Act.

There is no mention of his date of death.

Robert Hunter

State of New York

City and County of New York

On this eighteenth day of July in the year of our Lord one thousand eight hundred and thirty two personally appeared before the Court of Common Pleas of the City and County of New York, Robert Hunter, a resident of the City of New York in the State of New York who first being duly sworn according to Law doth on his oath make the following declaration in order to obtain the benefit of the provision made by the Act of Congress passed June 7th 1832. That in the month of October in the Year one thousand seven hundred and seventy five at the City of Philadelphia in the State of Pennsylvania he enlisted as a Sergeant of Marines on board the Brig Andrew Doria, commanded by Captain Biddle – That in the month of December in the same year he sailed on board said Brig from the Delaware River, the said Brig constituting a part of the fleet commanded by Commodore Hopkins, they proceeded on a cruise to the West Indies and from Nassau in the Island of New Providence, they brought off a quantity of heavy cannon & stores & returning to the Continent they were chased into New London and blockaded by the British Fleet – and in the year one thousand seven hundred and seventy six the Marines were discharged from further service.

Author's note – Robert Hunter's pension application goes on to detail further land service.

His pension application was approved and he received $418.32 per year under the1832 Pension Act. All his credible service was documented as being an Ensign, Lieutenant, & Captain in the New York State regular army. His se service was not credited.

He passed away on the 8th of May 1835.

Daniel Harper

District of Pennsylvania, Mercer County

On this twenty sixth day of May 1819, before me the subscriber, one of the judges in and for said county. Personally appeared Daniel Harper aged sixty five years, resident in Mercer County PA, who being by me first duly sworn, according to law doth on his oath, make the following declaration in order to obtain the provisions made by the late Act of Congress, entitled "An act to provide for certain persons engaged in the land and naval services of the United States in the Revolutionary War."

That he the said Daniel Harper in May 1775 entered at New York as Boatswain in the schooner Liberty under Comodore

Smith on Lake Champlain and continued until October of November when he was discharged. In the same year he shipped as gunners mate with Comodore Hopkins at Philadelphia.

Author's note – It is not stated in his initial claim, but is later corrected in subsequent correspondence to state that he shipped aboard the *Alfred* with Commodore Hopkins.

They sailed next spring and took the island of New Providence where they got 74 pieces of cannon, two magazines with many other things. In April 1776 they took a small vessel of eight guns and that night were attacked by a large ship of war. We engaged her 5 hours but she run from them in the morning. They put in to New London and refitted. They then went to providence & landed their cannon. Then said Harper with about 20 other went aboard Andrew Doria, Nicholas Biddle of Philadelphia, in June or July they took two ships having 250 highlanders aboard of each ship. Soon they took a ship with 400 hels of sugar, 10 pemcheons of rum, 1 ton of Black Ebony on board. They then took a ship from Norfolk with some dry goods aboard. They soon after fell in with and discharged several of Dunmore's fleet

Author's note – The spelling is bad and difficult to read in this claim, and I transcribe as written – or at least as best I can figure out. I'm not sure what measurement he is trying to describe for sugar and rum. Also, Dunmore was a an influential British sympathizer from Virginia.

I returned to Philadelphia in the fall. They refitted and Isiah Robeson took command of the vessel, which said Harper shipped and received a warrant as gunner. They sailed to St. Statia

Author's note – I think he is referring to St. Eustatius on the Island of Oranjestad, part of St. Kitts & Nevis.

And returned in March to Baltimore when he shipped aboard the Defence, 26 guns, Captin Cook, as Boatswain. Being in all betwinst two and three years in the Continental establishment. And quit the service at Baltimore.

Author's note – Daniel Harper's pension application was approved and he received $8.00 per month under the 1818 Pension Act.

There is no mention of his date of death.

USS Alfred

The *Alfred* was a converted 140' merchant ship with 20 to 24 guns and a crew of around 200 men. She was acquired in 1775 and sailed under the United States flag until she was captured by the British in 1778.

Her captains were said to be Dudley Saltonstall, John Paul Jones, and Elisha Hinman.

Alfred - United Staes Navy.

Charles Bulkeley

State and District of Connecticut

New London – December 30 1818

On this 30th day of December 1818 before me the subscriber & Judge of the Superior Court for said state of Connecticut personally appeared, Charles Bulkeley, of New London in said state and being by me first duly sworn under law did on his oath make the following declaration in order to obtain the provision made by the late Act of Congress entitled an "Act to provide for certain persons engaged in the land & naval service of the United States in the Revolutionary War."

That in the year 1776 he entered at New London aforesaid as a midshipman on board the United States ship Alfred, Dudley Saltonstall Esq Commander in Commodore Hopkins Squadron -That soon after in the same year he entered on board said ship and sailed from Delaware and proceeded to New Providence.

When they took New Providence - after which he resumed in said ship to New London where they lay some time – when they proceeded to Providence River in Rhode Island – where she also lay for some time – while in said Providence River in the year 1777 he was appointed sailing master of the Brig Hamden – But by permission of Commodore Hopkins continued on board the Alfred- soon after which John Paul Jones Esq took the command of the Alfred at Newport. That he proceeded in said ship under John P Jones in company with the US Sloop Providence, Capt. Harker on a cruise off Cape Briton where they took many prizes – That he was put on board one of the prizes as Prize Master and arrived in her at Bedford – The ship Alfred returned to Boston where he soon after again joined said ship Alfred when Elisha Hinman Esq took command of her. At which time he was appointed a Lieutenant by Commodore Hopkins in which capacity after the ship was repaired – he sailed in her under said Hinman from Boston & proceeded to Portsmouth where they joined the United States ship Raleigh, Capt. Thompson, and company with the Raleigh sailed for France – In the course of the voyage took two merchant men and proceeded to L'Orient – that they remained at L'Orient for several months during which time they repaired the ship – when they were ordered away – after leaving L'Orient, they cruised off the Madeira Islands, Cape De Verde & Coast of Africa and took one prize. From thence proceeded to the Windward Islands in the East Indies – while at the Windward Islands were captured in the year 1778 by the ships Ariadne & Ceres and carried to Barbados – where they were detained until the convoy for England was ready for sailing – when the officers of the Alfred, Elisha Hinman master, Lieutenant Peter Richards and himself – together with John Welch Captain & Hamilton Lieutenant of marines were put aboard the Yarmouth, an old seventy four resized to a sixty four, and sailed as prisoners to England – Where they arrived & were confined in Falkland Prison – After being there confined for some time, they made their escape to London & from thence to France. After remaining in France for some time he took passage from Bordeaux to Baltimore and off the Capes of Virginian he and Lieutenant Richards were put on shore about ten leagues to the southern end of Cape Henry then on to Boston, where they arrived in 1779.

Author's note – Charles Bulkeley's pension application was approved and he received $20 per month for one year service as a Lieutenant under the 1818 Pension Act. He reapplied under the 1832 Pension Act and received $360.00 per year for his service as a Lieutenant.

There is no mention of his date of death.

John Frisk

United States Of America

Common Wealth of Massachusetts, County of Worcester

On this fourth day of September AD 1832 personally appeared in open Court being a court of Probate within aforesaid County and Court of Record now in person in said County – John Fisk a resident of Northborough in the County aforesaid and Commonwealth aforesaid aged 72 years who being first duly sworn according to law doth on his oath make the following declaration in order to obtain the benefit of an Act of Congress passed June 7th, 1832- that he entered the service of the United States under the following named officers and served as follows –

This applicant says that he was born in Cumberland in the State of Rhode Island in the year 1760 that his birth is recorded in that town that he lived in that town at the time he entered the service & since the Revolutionary War lived there until 1794 when he removed to Westborough in said County of Worcester where he lived seven years, he then removed to Northborough where he has lived ever since and where he now lives.

And this applicant further says that in the year 1775 in the month November he enlisted as a private for one year in Capt. Abimelech Riggs Company & in a Regiment commanded by Col. William Richmond a Regiment of Rhode Island State Troops raised for the defense of that state & of the United Colonies and served in said Corps until November 1776 when being honorably discharged. He soon afterwards entered as a marine on board the Frigate Alfred a vessel of war in the service of the United States commanded by Paul Jones, Esquire. He entered on board said Frigate at a placed called "Holm's Hole" soon afterwards we sailed on a cruise. We cruised off Halifax and took from the enemy seven prizes, one of which was very valuable having on board eleven thousand suites of clothes besides a quantity of arms & brass field pieces. We arrived in Boston from said cruise in May 1777 when he was discharged & sent home.

Author's note – John Frisk's application goes on to detail several other times he enlisted or was drafted for military land service.

John Frisk's pension application was approved and he received $66.00 per year for 20 months as a private under the 1832 Pension Act.

There is no mention of his death in his pension application.

Richard Grinnell

Richard Grinnell's Declaration

On the 27th day of April AD 1818 Richard Grinnell aged 66 years a citizen of the United States and an inhabitant of the town of Union in the County of Lincoln in the state of Massachusetts makes the following declaration - That sometime in the month of March AD 1776 he entered on board the United States ship of war Alfred at Philadelphia commanded by Capt. Saltonstall in the fleet under the orders of Commodore Hopkins – That he sailed & lived on board said ship nine or ten months & was then transferred to the sloop of war Providence where he served four months, making his whole term of service in the navy more than a year at which time he actually and faithfully served this country against the common enemy & on the Continental establishment and at the close of service received an honorable discharge which was burnt with the house in which he afterwards lived. That he was discharged at Providence in the state of Rhode Island – that during his said term of service he was at the taking of New Providence – and was also in an engagement with the Glasgow Frigate – and also helped to take a Bomb Brig from the enemy

Author's note – Richard Grinnell's pension application was approved and he received $8.00 per month for one year of service from the 1818 Pension Act.

There is no mention of his date of death.

James Palmer

State of New York

County of Livingston

On this 24th day of September in the year 1832 personally appeared in open court before the Judges of the Court of Common Pleas of Livingston County now sitting, James Palmer, of the town Mount Morris in the County of Livingston & state of New York aged seventy eight years who being first duly sworn according to law doth on his oath make the following declaration in order to obtain the benefit of the Act of Congress passed June 7th 1832. That he entered the service of the United States under the following named officers & served as herein stated-

In the month of August 1775, he then being then twenty one years old residing in the town of Stonington New London County Connecticut – the village of Stonington was attacked by the British ship Wallace. After the ship went off troops were raised for the purpose of fortifying & defending the village when he enlisted for one month. After the month had expired a regiment was raised under the command of Colonel Webb, when he the said Palmer enlisted in the company of which Oliver Smith was Captain – Nathan Palmer First Lieutenant & Robert Staunton Second Lieutenant. The company remained at Stonington till the fort was finished & were then ordered to New London to aid in building Mamacock Fort & there remained until the first of December of that year. He then enlisted as a sailor on board of the Continental Fleet under Captain Elisha Hinman. After remaining a short time at New London he sailed with others for Philadelphia. The river being frozen over they rendezvoused at Cohansie until the fleet came down in the spring and the said Palmer went on board the ship Alfred, which was the Commodore Ship, Ezek Hopkins was Commodore – Dudley Saltonstall was Captain of the Alfred – Paul Jones First Lieutenant, & one Sebra Second Lieutenant. The fleet then sailed for New Providence Island and the crew landed on the south side of the island & marched & took the south fort & afterwards the city- & so much of the military stores of the island as the fleet could carry away – they also took & carried away Governor Brown as prisoner of war.

While they were lying at the island, they took a ship belonging to Governor Brown which was afterwards ransomed. On their return they fell in with the British fleet off Montauk Point. They first captured a brig called Bomb Brig - & a schooner belonging to the British fleet. Afterwards fell in with the ship Glasgow & commenced a battle with which continued until two British frigates hove in sight. They then left the Glasgow & sailed for New London, where they landed their sick & wounded. While lying there General Washington passed through on his march from Boston to New York. After the fleet was repaired they sailed for Providence & lay there until they could make preparation for another cruise. While they were lying at Providence all the officers were summoned to Washington to answer for some misconduct & were cashiered. After the crew had remained some time on board the fleet it was ascertained that the Continental agent at Providence whose name was Nightingale as he the said Palmer Believes – had power to discharge the crew – When he applied for & obtained a discharge.

Author's note – James Palmer goes on to detail his land service after being discharged from the *Alfred.*

His pension application was approved and he received $60.00 per month for seven months as a private & seven months as a sailor from the 1832 Pension Act.

He passed away on the 15th of September 1840.

Nathaniel Richards

I Nathaniel Richards of the town & county of New London & State of Connecticut do depose & declare that from a time in the spring of the year 1777 I was appointed a Lieutenant of Marines on board the U.S. Ship Alfred, Elisha Hinman Esqr Commander. In my appointment I repaired on board said ship in the port of Boston, after remaining there for a month I sailed in said ship on a cruise when we captured several prizes and proceeded with the same to for L'Orient in France where we spent most of the winter. Thence we sailed on a cruise off the Island of Madeira, thence to the coast of Africa – thence to the West Indies and a battle to windward of Bahamas and was captured on or about the 10th day of March 1778 by two of his British Majesties ships the Ariadne & Ceres commanded by Captains Pringle & Dakew. On our arrival at Barbadoes after capture, about the 26th of this same month, I was recognized by Capt. Thompson of the British ship Yarmouth a 74, who knew me when I was a child & was intimate in my Fathers family many years before. By his influence with Capt. Pringle of the Ariadne and by the intervention of Capt. Hinman, I was permitted to return home on parole.

Author's note – During the Revolutionary War both sides paroled enlisted & officers. This custom allowed men to return home as long as they agreed not to fight again, unless they were exchanged for other prisoners - which was common.

Capt. Hinman was particularly desirous for me to return that I might give an account as early as possible to the navy board of the manner of our capture and of the disposition of our consort the Frigate Raleigh, and in order to affect the object of his request, represented the fact of my discharging the duties of Purser of the ship as well as those of Lieutenant of Marines.

From Barbados I proceeded on a Corvette for Martinique where I took passage for America in the Brig Charming Lady, E. Hammond master. I was again taken by his Majesties Frigate Ambuscade, McCartney Commander – carried into Halifax – and imprisoned. Hence, I was released and traveled in a Corvette the 7th day of July 1778 arriving at New London my native place the 28th day of the same month, after an absence of about fourteen or fifteen months. On my being exchanged, which as well as I can recollect was about the first of October 1778, I received the appointment of Purser for the U.S. Frigate Confederacy, Seth Harding Esq. Commander. This appointment I continued to hold until the capture of said frigate by a British Squadron I believe in the month of April 1781

When we sailed from Boston in the Alfred they lacked an officer as Purser, I believe this was not known to the navy, nor have I any distinct recollection at this time that any anyone was on board, but on our arrival in France, many of the crew being without clothing, a quantity. was furnished by Capt. Hinman. At his particular request I attended the purchase of them and made the necessary entries and did other duties in the line of a Purser and I was requested to perform this additional service. I believe no compensation was ever made me. I think no provisions of the ship came under my charge, but only clothing. When the rolls of the officers & crew of the ship Alfred were made out for settlement, my name could not be entered on both capacities and I can not recollect if I was entered as Lieutenant or Purser.

Author's note –Nathaniel Richard's pension application was approved and he received $20.00 per month for eighteen months as a sailor from the 1818 Pension Act.

He passed away on the 1st of June 1832.

John Brinckly

I John Brinckly of Charleston in the County of Middlesex & Commonwealth of Massachusetts Doth on oath declare that during the Revolutionary War between America and Great Britain in the month of July in the year of our Lord One Thousand Seven Hundred and Seventy Seven I ship'd as Carpenter's Yeoman on board the ship Alfred of 20 Guns, Capt. Elijah Hinman, sail'd from Boston to Portsmouth N.H. in August and thence in company with the Rawleigh, Capt. Thompson, about the first of September we sailed on a cruise – captured a ship & brig and carried them into L'Orient in France, laid there about two months to repair, sailed from thence and on the cruise touched at Si. Sago Cape de Verde for water, thence along the coast of Africa, touch'd at Gorer, thence on a cruise was captured by his Majesty's ship and sloop of war Ariadne & Ceres off Barbados where we were carried in and imprisoned about a week, then got over to Martinico and then sail'd in the Brig Resistance 18 guns for Boston and arrived as I believe in May 1778. I continued to serve on board of her until July of the same year.

Author's note – John Brinckly's pension application was approved and he received $8.00 per month for nine months as a sailor from the 1818 Pension Act.

There is no mention of his death.

Gurdin Burnham

I had contemplated skipping right to Mr. Burnham's tour on the *Alfred,* because his claim was long. But it was so well written, interesting, and detailed that I have included his entire pension application.

State of Illinois Eager County

On this third day of June 1833 personally appeared in open court being a court of record to whit The County

Commissioners Court in for the said County of Edgar, Gurdin Burnham, a resident of the County of Edgar & State of aforesaid aged 77 years on the 13th day of February last who being first duly sworn according to law doth on his oath make the following declaration in order to obtain the benefit of the provision made by the Act of Congress passed June 7 1832. He states that he was born in East Hartford in the State of Connecticut on the 13 day of February 1756. This his age he gives from the family register kept by his father. That in the last mentioned town and state about the 12th or 15th of May 1775 he enlisted for 8 months as a drummer under Captain George Pitkin & was marched to Roxbury in the State of Massachusetts. Thence to Brooklyn in state last mentioned; Captain Pitkin now commanded as Colonel. In June was marched in Colonel George Pitkin's Regiment to a certain high hill near Roxbury & in about three weeks to the south of Boston in past named state to a place called Squantum Farm, a certain Captain Parsons belonging to the regiment with his company was stationed at a place called Hingum. This applicant under the command of Lieutenant Holdridge (or Holdrich) commanding as captain was stationed at Cohassel where he continued till the expiration of his said 8 months service on continental establishment. When & where he instantly again enlisted under the said Captain Holdridge for the term of 12 months on continental establishment & marched to Roxbury, placed under the command of Colonel Thomas Willis & remained at Roxbury until 17th March 1776 about which time the British left Boston & marched into it, & about 8 days after returned to our barracks at Roxbury, & about 12th of

April marched to New London in the State of Connecticut & there embarked on vessels & sailed for New York where we arrived on the 4th May 1776. & in about 3 weeks we marched & began to build Fort Washington about 9 miles from New York, in about 7 days returned to New York & pitched our tents at the Jews burying yard. While remaining there proclamation was repeatedly made & read to the lines that any men who would undertake to burn the following British vessels, the Phoenix a 40 gun ship, the Rose a 20 gun ship, the Thunder bomb & two Tenders here lying in Tappan Bay, should receive $1000 each & be discharged from service if requested. That pursuant to said proclamation he joined in said enterprise & went on board a fireship commanded by Captain Thomas, which in company with another fireship commanded by Captain Halbert, moved off from Albany pier in East River & steered up the Cheval de frise & anchored again.

Author's note – A Cheval de fries is a defensive obstacle, in this case, placed in the water to prevent passage.

Next day there came on a heavy thunder gust & wetted us so much that himself & 4 men were sent back to New York to get new priming & some provision, especially a demi John full of rum for Captain Thomas

Author's note – A demi John is basically a large jug. In this case, a jug of rum.

By his order on our return to the fireships he (this applicant) & one Harris cleared & primed the fireships a new except two barrels & powder in the hinder part of the ship, which prevented our setting fire to the matches in the after part of the ships, but fired. Then Captain Thomas got intoxicated & went over board & was afterwards found dead, supposed to have been killed by a cannon while in the act of swimming to the shore. In firing the ships this applicant was so situated that the boats provided to take them to shore left him naked & severely & powerfully burnt by the flash of powder explosion, had to swim to shore a distance of two miles, that he was taken up in his awfully burnt situation & carried to Gen. Washingtons quarters to whom he gave a statement of the transaction in burning the ships – and by whom he was ordered to be conveyed to his father's house at East Hartford where he remained till recovered of his wounds, which was about the first of November following, during which time he made application to Ralph Pomeroy – Paymaster General for his part of the fireship money & received answer that the fire ship money had not come to hand, nor did this applicant ever receive all or any part of said fireship money.

That about the 12th of June 1777 he entered on board the ship Alfred of 28 guns, Elisha Hinman Commander, Peter Richards 1st Lieutenant, Deval 2nd Lieutenant, Brickley 3rd Lieutenant, Hamilton Captain of Marines, Cockrell Boatswain, and about the last of August sailed from Boston to Portsmouth in the State of New Hampshire & fell in with the ship Raleigh commanded by Commander Thompson & in about 10 days both ships set sail on a cruise for France, on our outward bound passage fell in with & captured two British vessels bound for Halifax in Nova Scotia, plundered them of 4 reams of continental paper money struck by Riringtors the Kings printer in New York, not signed, the prisoners we took on board & burnt their vessels & paper money. In 8 or 9 days fell in with a large Barque bound from Jamacia to London captured he, put a prize master on board & sent her to North America. After sometime we arrived in France where we were detained about 4 months repairing our vessels. Thence sailed thru the Bay of Biscay & the wine islands thence to the Barbary Coasts to Senegal Town to Gurak, to the Cape De Verde Islands & the Isle of May, there took in upwards of 200 goats for the use of the ships crew, water at St. Gago, or Yoga, sailed to the West Indies & about 3:30 to the windward of Barbados fell in with two British ships of war. One a 28 & the other an 18 gun ship & an engagement ensued in which the powder chest of the vessel on which this applicant was blown up, killed 12 men & burnt this applicant almost to death. We were all made prisoners & carried to Bridgetown in Barbados & there put in jail 81 in number I think. This applicant was so badly wounded and burnt that he could not walk one step after being in prison 62 days. Was put on board a vessel & carried to Martinico & there exchanged & in about two weeks put on board of a continental vessel which had been repaired by our agent there & sailed for Boston in the state of Massachusetts where we arrived in about two months.

In the last of August or 1st of September 1778 he then returned home to his fathers & remained there till the summer of 1779 when he again entered into service on board the Brig Nancy commanded by William Burke & sailed to pilot the French fleet under Count De Esting from Newport to Boston, in which service the Brig Nancy was captured by the British ship Oraile Commanded by O'Brian. We were all ironed, put aboard a British ship, carried to Lord Howes Fleet & distributed among the vessels, & after awhile sent to New York & all put aboard the British prison ship Felicity & about 40 days was put on board a Carteil & sent to Boston where paroled for 40 days in which time we was exchanged.

Author's note – Gurdin Burnham's pension application was approved and he received $85.77 per year for nine months as a sailor under the 1832 Pension Act.

There is no mention of his death.

James Cassell

Commonwealth of Massachusetts,

Suffolk, Boston

On this twenty eighth day of August A.D 1832 personally appeared in open Court before the Municipal Court of the City of Boston, now sitting at Boston within and for the County of Suffolk aforesaid, James Cassell, of said Boston aged seventy four years. Who being duly sworn according to law, doth on his oath make the following declaration in order to obtain the benefits of the Act of Congress, passed June 7th, 1832. That he entered the service of the United States, under the following named officers, and served as herein stated. That is to say that in or about the month of September 1777 he shipped at Boston in the 20 gun ship Alfred, Captain Hinman, in the service of the State of Massachusetts. He sailed in her to Kittery in Maine, where they joined the Frigate Rolla, Captain Thompson.

Author's note – the *USS Raleigh* is frequently referred to as Rolla in these pension claims.

Sailed from thence on a cruise, touched at L'Orient in France, thence to the coast of Africa, where they took several prizes near the mouth of Senegal & after being out about 16 months were captured by the Ceres & Ariadne – British ships of war, & carried into Barbadoes. He was detained some time in captivity, was at length released & proceeded forthwith to Boston, which he reached about April 1779 after a service of about nineteen months to the best of his recollection. Soon after this, the exact date he does not recollect, he shipped again at Boston in the Frigate Rolla, commanded by Commodore Berry & sailed on a cruise; soon after leaving port was pursued by the British 50 gun ship Experiment, after a severe conflict run ashore on Fox Island where the Rolla was dismasted & taken. This deponent with the rest of the crew escaped in the Barge & returned to Boston. He was engaged in this service about three months.

Soon after this, the exact period he cannot now recall, he shipped again at Boston in the Brig of War Resistance, then in the service of the State of Massachusetts, sailed on a cruise to the West Indies, took several prizes & returned to Boston after an absence of from two to three months as near as he can recollect. His wage were eight dollars per month.

Author's note – James Cassell's pension application was approved and he received $8.00 per month for 9 months of service under the 1818 Pension Act. Later, he re-applied and received $96.00 per year for two years as a sailor from the 1832 Pension Act. For some reason this was reduced in 1842 to $52.00 per year for 13 months of service.

I chose to use his 1832 Pension Application over his 1818 claim because his 1818 claim was a mirror copy of John Brinckly's, who you read earlier.

There is no mention of his death.

Stephen Northrop

District of Rhode Island. Washington County

On this 31st day of August in the year 1819 before me the Subscriber one of the Judges of the Court of Common Pleas for the County of Washington in said District personally appeared

Stephen Northup aged 66 years resident in the town of South Kingston in the said County & District aforesaid who being by me first duly sworn according to Law, doth on his oath make the following declaration in order to obtain the provision made by the late Act of Congress entitled "An act to provide for certain persons engaged in the Land & Naval services of the United States in the Revolutionary War;" That the said Stephen Northup enlisted in the States service on the 2nd day of August 1776, the said Steven Northup enlisted under John Paul Jones, Commander of the Continental ship called Alfred in the United States Navy. That he enlisted as aforesaid in Newport in the District aforesaid for the cruise which was to the English Channel. That he was in seven battles or engagements taking seven prizes, 5 ships 1 Brig, and a Snow.

Author's note – A snow is a square-rigged vessel with two masts complimented with a snow, or trysail mast stepped behind the main mast.[2]

Two of the prizes the Ship Malice & John were carried into Boston, the Ship Alfred as aforesaid was gone about 10 months in said cruise in which said Stephen Northup continued to serve on board of said ship in the service of the United States until the return of said Ship which was for more then the term of 9 months.

Author's note – Stephen Northup goes on to detail land service on several occasions.

His pension application was approved and he received $8.00 per month for ten months as a sailor from the 1818 Pension Act. None of his land service was credited.

There is no mention of his death.

Letter From The Commonwealth of Massachusetts

Portsmouth, September 27, 1777

By a letter from Captain Thompson of the Continental Frigate Raleigh we are informed that they were all the 8th instant in lat. 41.30 long. 46.30; was then in pursuit of a windward island fleet under convoy of three men of war: one of which she engaged for 45 minutes and had obliged them to quit their quarters: 15 minutes before she left her, only one man was left on deck, she was in a shattered condition and no doubt lost many men. The loss on our side is trifling, only one man killed. There is no manner of doubt but their whole fleet must have fallen into our hands provided the Alfred could sail equal to the Raleigh.

Portsmouth April ,7 1778

Yesterday arrived here Captain Thompson in the ship Raleigh of 32 guns; he sailed from hence last August and has during the cruise taken six prizes, five of which got safe into port. He informs that the Alfred, who sailed with him is taken.

Boston April 9, 1778

The Raleigh and Alfred sailed from L'Orient the 29th of December in company and proceeded to cruise off the coast of Africa; from thence crossed the Windward Island and from that to Boston. Saw no English vessel during the cruise accept one sloop with wines which they took off Senegal Bar from an Anchor- until they come to the windward of Deseada where they fell in with two British ships of war. The Alfred bore away at a very unfavorable time, to try to make her escape and was taken.

Boston Thursday May 7, 1778. Saturday returned into port a privateer brig of 14 guns, lately commanded by Captain Chew of Connecticut, who was killed in an engagement with a twenty gun ship. After which the brig made off and got safe into Martinique where she refilled; and from whence she sailed thirty-eight days ago with a number of Capt. Hinman's men, 21 of whom died on their passage.

Author's note – This letter was found in Ebanezer Leman's pension application. I did not include his declaration because it was filed by his wife who, was seeking a half pension form the 1834 Pension Act, and not in his own words. There was however a letter form James Cassell, a shipmate of Ebanezer's stating he served with him on *Alfred.*

USS Baltimore

The *Baltimore* was a 12-gun brigantine purchased for the Continental Navy in 1777. She sailed under the flag of the United States until she sank in a storm in 1781

Her captains were said to be Thomas Read.

James Hays

State of Pennsylvania, city of Philadelphia. In the Court of Common Pleas in the court of Philadelphia on this twenty fifth day of April Anno Domini 1833 personally appeared James Hays of the city of Philadelphia and state of Pennsylvania, aged seventy eight, who being first duly sworn according to Law doth on his oath make the following declaration in order to obtain the benefit of the provision made by the act of Congress passed June 7 1832.

That in the year 1776 he lived in the city of Philadelphia where he had been born. In that year he joined as a volunteer a company of artillery of Pennsylvania troop in the service of the United States commanded by Captain Stiles, the company was raised at Philadelphia – it belonged to the 3rd regiment of state artillery, Colonel Eyre – he marched in the summer of 1776 with his company, he being a private, to Amboy in New Jersey – where he was engaged in camp duty in the neighborhood of the enemy for two months and upwards – they were marched back to Philadelphia and this declarant was honorably discharged from active service, still continuing a member of the company – In the next year, he went with the same company under Captain Fry (who had been first lieutenant at Amboy) down the Delaware to Billingsport New Jersey – they were driven from there after some weeks by the British troops. And went to the Pennsylvania side of the river to Fort Mifflin – after some days they were marched to Head Quarters White Marsh, Montgomery County Pennsylvania north of Philadelphia – they were camped there and remained till about Christmas when they were discharged – The British being in Philadelphia – the declarant went up to Allentown Pennsylvania and worked at making cartridges in the laboratory, then under charge of his old Captain Stiles – From the time he went to Billingsport till he was discharged at White Marsh it was four months at least – during which time he was in military service.

After the British left Philadelphia, he returned to the city and worked for a time at his trade as a carpenter – He then entered as a carpenter on board of the United States Brig. Baltimore, a packet, - Captain Nicholson, then fitted out at Philadelphia for the purpose of convoying ships with provisions for the French fleet at Cape Francis -

Author's note – Cape Francis is in Haiti

The Baltimore carried ten, four pounders, she sailed with her convoy of 5 vessels and arrived at Cape Francis with two of the vessels, the other 3 being lost in a gale and never heard of again – after repairing the effects of the gale on the Baltimore, and remaining some time in the West Indies – they returned homewards, and were castaway on Cape Henry in 1781 and lost everything –

Declarant after much privation and suffering reached Philadelphia by land – He served the United States in the Baltimore full eight months.

After he came home he joined an artillery company and was called into active service for two months – The company was a volunteer corps and commanded by Captain McCulloch and belonged to Colonel Marsh's regiment of state troops in serving the United States – they were sent down to Billingsport and served in barracks full two months, when they came home and were discharged.

He says that in addition to the services already mentioned he went as a volunteer with Commadore John Barry on an expedition up the Delaware towards Trenton in boats – It was after he had been at Amboy and in cold weather when the Hessians were at Trenton. He served on this occasion five or six weeks.

Author's note – James Hay's pension application was approved and he received $62.66 per year for 9 months service as a seaman in the navy and 8 months service as a private in the army.

There is no mention of his date of death.

In an interesting affidavit from Jacob Wayne, who was a character witness to James's service, Mr. Wayne states that he witnessed the sinking of the *Baltimore Packet* and the crew swimming ashore. Mr. Wayne was himself on the vessel *Mars* and she sank soon after the *Baltimore Packet*.

Just a side note, Mr. Wayne states that the *Mars* was chased into St. Eustatius, a Caribbean Island, by the British. There they took on cargo and were headed to the Chesapeake Bay. Knowing that the *Baltimore* was armed, the *Mars* more than likely fell in with her on her way to the Chesapeake Bay for protection. That would explain why they were traveling together when they were sunk. He ended up on shore with James Hays and the other survivors of the shipwrecked crews. Together they walked to Portsmouth, Virginia.

Cyprian Barnard

State of Connecticut, County of Hartford

On this 20th day of September 1832 before the Probate Court for the District of Hartford in the County of Hartford and State Of Connecticut being a Court of Record, having by law a Clerk and Seal, personally appeared Cyprian Barnard a resident of Hartford in the County of Hartford and State of Connecticut aged eighty years, who being first duly sworn according to law, doth on his oath make the following declaration in order to obtain the benefit of the Act of Congress, passed June 7, 1832

That he entered the service under Joseph Trumbull in the Company Department (said Trumbull was then Commissary General U. States in April 1775 at Cambridge, five days previous to the Battle of Bunker Hill and continued to serve under said Trumbull as deputy commissary until 1779 at the rate of forty dollars per month and was employed in purchasing and providing for the army – and procured furniture and other articles for the house of General Washington at Cambridge by order of the Commissary General. For proof of his service, the deponent has no documentary evidence and knows of no person now alive who can testify to the same. But would respectfully refer the Hon. Sec. of War to certain papers submitted by the deponent to the War Department at Washington in the year 1819 by the hand of the Hon. John Russ, then member of Congress from the State of Connecticut. & in 1824 Hon. Elisha Phelps, he thinks.

That he enlisted aboard the "Baltimore Packet" so called Sloop of war belonging to the U. States in 1781 or 1782, in July or August (according to his present impressions). Said Sloop carried 18 guns and was commanded by Commodore Thomas Reed of Philadelphia – That he served on board said Sloop of war five months and was discharged at Philadelphia – He served as a seaman. & we met & exchanged a broadside with the British Sloop of war Belvidere and should have taken her had she not been a better sailor & sheered off.

And dependent further declares that on his passage home from Antigua he was pressed on board the United States Frigate Trumbull at sea, but hired a substitute to take his place, being want to return home to his friends having been confined at Antigua as aforesaid. This took place in the month of June 1782 – according to his present impressions.

Author's note – For most of the war there was mandatory service for every male over the age of 18. It was common to hire someone to take your place instead of having to enlist due to the draft.

That he was drafted at Hartford in said state of Connecticut about 1779 and served as a private in the company commanded by his father, Captain Ebenezer Barnard, and was absent two or three months – and was discharged at Newton.

That he again entered the service as a drafted militia man with the same company commanded by his father, Captain Ebenezer Barnard & marched to Peekskill, then to Poughkeesore about September or October 1777 – but does not distinctly recollect the month – he was absent several weeks.

Author's note – Cyprian Barnard 's pension application was rejected and he was not granted a pension.

There is no mention of his date of death.

Jacob Smith

State of Rhode Island and Providence Plantations

County of Newport

On this _____ day of September A.D. 1832 personally appeared in open court before the Justice of the Supreme Judicial Court of said State now sitting at Newport, within and for the County of Newport, Jacob Smith, a resident of said Town and County of Newport and state aforesaid, aged seventy two years, who being first duly sworn according to law doth on his oath make the following declaration in order to obtain the benefit of the act of Congress passed June 7th 1832. That he entered the service of the United States under the following named officers and served as herein;

That he was born in Providence in said state, which place and Newport have been his domicil all his life, and that he is now seventy two years of age. That the following is a just and true narrative of his Naval service during the revolutionary war.

That your memorialist sailed from Providence on board the armed Sloop Montgomery of ten guns and eighty men in 1776, three cruises. First with Captain Daniel Bucklin, the same year with Captain William Rhodes, and in the same year Capt. Rootenburough, in which three cruises we made fourteen prizes. In 1777 your memorialist entered on board the Brig Fanny of fourteen guns and ninety five me, Capt. Whittlesea, sailed from Providence and returned into New Bedford, in which vessel we made six prizes, and was on board said Brig ten months and cruised off the Coast of France and West India Islands. He then entered on board the ship Rattlesnake of Philadelphia mounting twenty guns and one hundred and forty men, commanded by James McColler, in which ship we made ten prizes, four burnt and six sent in and was on board said ship six months. He then went on board the Merchant Brig Bee, Capt. Sturgess, bound to France, off the Coast of France was taken and put on shore at a place called Chrochet near Nants.

Author's note – Nantes, France

Travelled to Nants and shipped on board the United States Brig Baltimore Packet, Capt. Read, mounting twelve guns and sixty men and in November 1778 arrived in Philadelphia. Was on board said Brig four months at twelve dollars per month. Sailed from Philadelphia in March 1779 In a Merchantman and was captured the next day off the Capes by the Brig Diligence and carried to New York. Ran away from New York and travelled to Boston were he entered on board the United States Ship the Queen of France, Capt. Rathbone, was on board said ship over six months at twelve dollars a month, in which time we took nine ships and a Brig. In company with the Ship Providence, Commodore Whipple, and Ranger, Capt. Simpson. In 1780 entered on board the Schooner Hibernia Privateer, Capt. O'Brien, of fourteen guns and eighty men in which we made four prizes, then went on Board the Brig Poggy, Letter of Marque of sixteen and sixty men as second officer, commanded by Captain Armstrong, was on board said

Brig three months at twenty two dollars a month, during which we took a large transport ship laden with stores bound to New York after a hard fought action. In the same year went on board the ship Port Packet, Capt. Ebenezer Stocker, of Newbury Port, of fourteen guns and forty men as second officer, bound to Saguia and thence to Spain. Was on board said vessel seven months at twenty six dollars a month. On our passage made one prize off St. Andora, fell in with an English Privateer Lugger of sixteen guns and one hundred men. Our ship had only fourteen guns and forty five men. After a hard fought action we beat her off with the loss of one man killed and three wounded, and arrived the same day in Bilboa. From thence went to Bayonne and entered on board the ship Alexander, Capt. Gregwor, of twenty eight guns built in Bayonne and commissioned by Doct.. Franklin, entered on board said ship Feb. 20, 1782 as sailing master at forty dollars per month. Sailed in April for Bordeaux to take in our armament, having only on board ten guns and sixty men, two days out fell in with the Brig Greyhound from Bristol England mounting twenty guns and one hundred and twenty men, fought her from twelve o'clock M until six o'clock P. M. and got into Rochelle with the loss of four men killed and twelve wounded, myself slightly. Sailed from thence to Bordeaux were we refitted and took in our armament. Sailed on the tenth of August to cruise off the West India Islands, was captured on the 17 Aug. by the Mediator of fifty guns, Sir James Sutterel, Master, off Cape Finnister

Author's note – Cape Finisterre is off of Spain.

and was carried to England and put in Mill Prison. Was kept there until the peace, and then sent to Merleaux in France in March, and travelled to Nants and went on Board Brig Providence and arrived in Providence Rhode Island the 21 July 1783, having been on board said ship and during the time your Memorialist was in prison and until his return to America seventeen months.

Your Memorialist having been during the whole of the war on board some armed vessel commissioned by the States, or by the Congress of the United States, and having been four times a prisoner, and at the taking of over sixty British Vessels.

Author's note – Jacob Smith's pension application was approved and after all that sea service he only received $25.00 per year under the 1832 Pension Act. He was only credited for 6 months of service as a seaman.

There is no mention of his death.

USS Bonhomme Richard

The *Bonhomme Richard* was a converted merchant ship with 42 guns and a crew of around 350 men. She was loaned to the United States by the French in February of 1779 and sailed under the United States flag until she was sunk by the British in September 1779.

She is famous for her Captain, John Paul Jones, uttering his famous reply when called to strike his colors "Sir, I have not yet begun to fight!"

Her captains were said to be John Paul Jones

Bonhomme Richard - from the National Archives.

John Hall

State of Pennsylvania, Berks County, on the 4th day of October 1850 personally appearing before me a justice of the peace in and for the county of Berks and state of Pennsylvania, John Hall, a resident of Leesport County and state aforesaid aged eighty nine years, who being first sworn according to law doth on his oath make the following declaration in order to obtain the benefit of the act of congress passed June 7th 1832;

That he went on sea, the service of the United States under the following named officers as herein stated; went to sea in the year 1779 in the month of August on ship Ariel, I entered under Captain Courder at Philadelphia in the state of Pennsylvania as a sailor. The ship Ariel was loaded with some tobacco at Philadelphia, which we have unloaded at Lorient in old France.

And at the same place we was dispatched under Paul Jones, captain, an the ship by the name Good Man Dick, with a load of powter to take to Philadelphia for ammunition for the war. Our first lieutenants name was Herman Diehl and our second lieutenants name was Benjaman Lunds, sailor master Herman Stacy.

Author's note – Good Man Dick was a term, or loose translation used for Bonhomme Richard by her crew.

One morning on sea we was attackted by on English ship, the name of it I do not know. And the next day again we give them four brought sides and we received two brought sides. They have attackted us the second time. Fires was given on both sides, our men worked like brave soldiers and we have shed their masts off and took about thirty of them prisoner. We have took them along to Philadelphia, there they was locked up and what became with them afterwards I do not know, and the powter we loaded on our ship we unloaded at Philadelphia and took it to the magazine. The ship Good Man Dick belonged to the King of France, he has sent afterwards a fricket and men to take the ship Good Man Dick to France, and between Philadelphia and New York the fricket and ship was both taken by the English, our captain, Paul Jones was a good man; in did and further do depose and say that he was on sea about three years and that Captain Paul Jones give him his honorable discharge at Philadelphia in the state of Pennsylvania where he was called into the service.

Author's note - Unfortunately his folder is very small and only states that his claim was rejected. There is no reason for the rejection and no date of death given. And like a true sailor he signs his deposition with his mark, an "X."

Also, in his deposition Mr. Hall's time line is a bit off from the official record. He mentions *Ariel* first, but *Ariel* wasn't lent to the Continental Navy by the French until October 1780. So, his tour on her would have been after the *Bonhomme Richard.*

James Mc Kinzey

District of Pennsylvania

On the third day of April A.D. 1818, before me Richard Peters, Judge of the District Court of the United States, in and for the Pennsylvania District. Personally appeared James Mc Kinzey, who being duly sworn, deposeth and declareth, that during the war of the Revolution he served against the common enemy as a seaman and afterwards as a cook in the navy on the continental establishment; that about the month of ------- A.D. 1779 he enlisted at L'orient, in France, on board the ship Bon Homme Richard, fitted out there for the Continental service by Doctor Franklin, under the command of Captain Paul Jones; that he served on board the said ship nearly a year, and all together about four years, that in the engagement with the English frigate Serapis, the deponent lost his right leg, and after his recovery he was made cook; that after the engagement between the Bon Homme Richard and the British frigate Serapis, the deponent, with the rest of the crew, went onboard of the Serapis, their own ship having been sunk, and went into Holland, thence to France, where they were put on board the Arial, hired by Doctor Franklin, and returned to America, and shipped again as captain's cook on board of the Trumbull frigate, Captain Nicholson, where he served about a year, where he was taken prisoner, and remained so at New York, four or five months.

That he was exchanged, returned to Philadelphia, and shipped on board the Continental ship Duke Lusaune, Captain Green, and served in her as cook until the peace.

Author's note – Correct spelling of Duke Lusaune is Duc De Lauzun.

That by reason of his reduced circumstances in life, he is in need of assistance from his country for support, that he is a resident citizen of the United States.

Author's note – James Mc Kinzey's pension application was approved and he received $8 per month for 1 year as a seaman under the 1818 Pension Act.

There is no mention of his death.

John Kilby

The Affidavit of John Kilby a citizen of the United States of North America and now the Commonwealth of Virginia deposeth & sayeth that on the 6th day of August in the year 1776 this deponent shipped on board the Brigantine Sturdy Beggar of fourteen guns commanded by Capt. James Cambell in the town of Vienna, Maryland, Dorchester County, near the place where he was born. then went to New Bern, North Carolina where the Brig then lay. We then sailed on a cruise and on or about the 20th day of November same year, we fell in with a Glasgo Brig of 6 double fortified six pounders and after a small action of ¾ of an hour we captured her and on the 1st of December same year, we fell in with the ship Smincy Galley of 10 guns and after an action of ½ an hour we captured her, the said Kilby was put on board the prize ship & on the 9th day of the same month was captured by the Resolution of 74 guns commanded by Sir John Chandler Oglesby, and carried into Spit Head, England and after remaining on board the guard ship nearly four months was carried to Mazel Hospital and after going through a trail the then sitting Judge calling over all navies pronounced sentence in these words to wit – "You are all condemned for piracy and high treason on his majesty's high seas." We the ships crew were then sent up to Fortune Jail under a strong guard of soldiers where we lay in a cell of brick walls near twenty two months, after which one hundred of us were exchanged over to France. We arrived in the port of Lonlap, then went up to the city of Nantze where we remained eight days. A letter was drafted and sent by post over to Capt. John P. Jones, then lying in the port of L'Orient, requesting him send over an officer that we wished to ship on board his ship, the letter was sent, they sent over two officers (the sailing master and the gunner) about 33 of us entered and signed the ship Bonhomme Richard's articles for twelve months on or about the 1st of April 1779, after which we sailed on a cruise from Port of L'Orient, where the sip then lay, but before we sailed we all received Twenty French Crown from Mr. Mailon the Continental agent. The before named cruise is well known to the whole world and needs no explanation from me. I will only say on the 3rd of November we fell in with the Rapris of 44 guns,

Author's note – He is referring to *HMS Serapis*

commanded by Richard Pearson and after an action of 9 glasses, four hours of which were side by side, we captured her.

Author's note – A measurement of one hour was referred to as a "glass," in reference to a sand filled hour glass.

I was an officer, first a midshipman, then masters mate, then Lieutenant in the action. Our ship, the Bonhomme Richard sunk within 24 after the action, we took charge of the prize ship and arrived safe in the Lexel (Holland) where we lay along time, after which our Capt. was by Doct. Benjamin Franklin ordered to take charge of the Alliance Frigate of 36 guns, and the then commander Peter Landais, ordered to go to Paris. We at length sailed on a cruise, went into Croney (Spain) where we lay some small time, we again sailed on a cruise after which we arrived in the Port of L'Orient, where we first sailed from. Our Captain (Jones) went to Paris, and on his return our ship was all ready to sail for the Port of Boston (America) but by some means the former Captain of Alliance, Peter Landais got on board of the ship while Captain Jones and some of his principal officers were on shore, the reason why it happened was this; As soon as Jones took charge of the Alliance the two ships crews and officers could not endure each other as Landais was accused of disobeying the orders of Jones, the Commadore of the squadron and generally called a coward by all the officers and men of the crew of that good ship called the Bonhomme Richard. The officers of the Alliance to get Landais, their captain in the ship again, his officers got him on board while Jones was on shore; Landais when on board way'd anchor and went to sea. We at length arrived safe in Boston where Landais was tried and at trial gave himself praise.

Author's note – Commodore John Paul Jones in his flagship *Bonhomme Richard* oversaw a squadron of vessels, one of which was *Alliance,* commanded by Pierre Landais. Commodore Jones had relieved Landais of command of *Alliance* for numerous counts of insubordination & misconduct. Probably the most notable is *Alliance* firing on *Bonhomme Richard* while she was engaged with Serapis!

While Commodore Jones was in Paris receiving orders from Benjamin Franklin, Landais did illegally take command of *Alliance* and sailed her to Boston. There he faced a trial, was found guilty, and removed from the service.

I do say that my discharge is not a dishonorable one. I followed the sea in armed ships and off the West Indies was again taken prisoner by a sloop of war of 16 guns commanded by Lieutenant Parker with dispatches from New York to Admiral Hood. I was plundered of all my clothes, my discharge, and all my papers I had before, obtained by Capt. Jones, acting as an officer.

Some little time after we fell in with the L'Amidable Frigate of 38 guns, Capt. Alexander Hood, and was put on board of his ship as a prisoner. Some small time after I was put on board the Tarbay 74 guns, ordered into Kingston, Jamaica, in which place I remained very nearly four months on parole. Further I will say that for all the said service serving my country in Congress service, upwards of two long years, I never received but the Twenty French Crowns and one Dutch Ducat, nine shillings and six pence sterling, either in wages or prize money.

Author's note – John Kilby's pension application was approved and he received $8 per month for 1 year of naval service under the 1818 Pension Act.

There is no mention of his death.

Aaron Goodwin

Commonwealth of Massachusetts

York – On this fourteenth day of May A.D. 1818 before me the subscriber, Chief Justice of the Circuit Court of Common Pleas for the Eastern Circuit in said Commonwealth, personally appears Aaron Goodwin aged sixty four years, resident in Parsonsfield in said county of York, who being by me first duly sworn according to law doth on his oath make the following Declaration in order to obtain the provisions made by the late act of Congress, entitled "an Act to provide for certain persons engaged in the land and naval service of the United States in the Revolutionary War."

That he the said Aaron Goodwin in the month of April 1779 at a port in France, entered as a seaman on board the ship called the Bon Home Richard, in the service of the United States, under the Command of Captain John Paul Jones, and sailed on a cruise in said ship and during an engagement the Bon home Richard took his British Majesty's ship called the Serapis – The day after the engagement the Bon Home Richard went down, and Capt. Jones and his crew took possession of the ship Serapis and proceeded in said ship to Holland, whence he left the Serapis and went on board the United States ship Alliance under the command of John Paul Jones, and from Holland in the last mentioned ship went on a cruise and went into the port of Corunna in Spain, and from thence to L'Orient in France, where Capt. Landas took command of the ship Alliance & sailed for Boston in America, at which port he arrived in the month of August 1780 where he was discharged.

Author's note – Aaron Goodwin's pension application was approved and he received $8 per month for 15 months of naval service under the 1818 Pension Act.

There is no mention of his death.

USS Boston

The *Boston* was a frigate built for the Continental Navy and launched in 1776. She was 114' in length and armed with 24 guns. She sailed under an American Flag until she was captured by the British in 1780.

Her captains were said to be Hector McNeill & Samuel Tucker.

Boston - Rod Claudis

Joseph Herrington

I Joseph Herrington a citizen of the United States, now resident at Wiscasset in the County of Lincoln in the State aforesaid, do on oath testify and declare, that in the war of the Revolution in March 1776 I entered and was engaged in the land service of the United States on the continental establishment and served accordingly from that time to March 1777, being one year, & being the time of which I enlisted as a private soldier against the common enemy, without any interruption or absence. That I belonged to Capt. Lanes Company, Col. Varneem's Regiment, in the Massachusetts Line – That I left the service in New Jersey in march 1777 when I received a regular discharge, which I have lost.

As soon as I received the above discharge I returned home & immediately in March 1777 I enlisted & shipped on board the United States frigate Hancock of thirty six guns, commanded by Capt. Manley, & served as a marine on board said frigate cruising against the common enemy until the 7th day of July 1777 when said frigate was taken & I was carried a prisoner to Halifax – I remained in Halifax a prisoner about nine months, & then made my escape & got back to Boston in March 1778, having served in said frigate & in prison, one year.

I then immediately, the next day I arrived in Boston on my escape aforesaid, in March 1778, entered & shipped as a seaman & marine on board the United States frigate Boston of thirty six guns, Capt. Samuel Tucker commander. We went in her to France to carry our Hon. John Adams, our minister,

Author's note – John Adams would go on to become the second President of the United States.

& after the service was performed, we cruised against the common enemy. I served in said Boston frigate one year & a half as nearly as I can recollect, in the above service & was then coming home in one of the prizes, when we were taken & I was held on board a British Man of War six months. I was then put on board one of their prizes & we rose upon the crew & retook the vessel & brought her into Boston in February or March 1780, having served on board said Boston frigate & in capacity as aforesaid, two years.

Author's note – Joseph Herrington's pension application was approved and he received $8 per month for one year of service as a private under the 1818 Pension Act. None of his naval service seems to have been credited.

There is no mention of his death.

Benjamin Crowninshield

State of Massachusetts, Essex County

On this seventh day of August A.D. 1832 personally appeared in open Court before the Honorable Daniel A. White, Judge of the Court of Probate, Benjamin Crowninshield Esquire, a resident of Danvers in the county and state above mentioned, aged 74 years, who first being duly sworn according to law doth on his oath make the following Declaration in order to obtain the benefit of the provision made by the Act of Congress passed June 7 1832.

That about the year 1777 he was appointed a Midshipman in the navy and entered on board the Frigate Boston, Hector McNeil Esq. Commander, in the Continental service and served on board her thirteen months; the exact dates when he entered and when discharged he cannot recollect; while on board he was in the action between that Frigate and the British Frigate Fox, in which the latter was captured.

The term of service has been stated and proved in a form, Declaration now on file in the War Department under Act of 1818.

He further declares that in July 1779 he was appointed Lieutenant of the States armed Ship Black Prince, commanded by Nathaniel West Esq and sailed from Massachusetts on the expedition against the British on Penobscot River in the Fleet under Com. Saltonstall, the land forces being commanded by Gen. Lovell and Gen. Wadsworth. The enterprise was defeated by Sir Geo. Collier, the British Commander, and American Fleet blown up and destroyed. He served on this expedition about three months.

Author's note – Benjamin Crowninshield's pension application was approved and he received $114.00 per year for 11 months of service as a midshipman & 3 months as a lieutenant under the 1832 Pension Act.

There is no mention of the exact date of his death, only that he passed in 1836.

Joseph Wilkinson

This may certify to whom it may concern that the bearer hereof – Jospeh Wilkinson, entered on board the Boston Frigate under my command in January seventeen hundred & seventy eight and continued with me nine months; acting in the capacity of ships corporal; about the last of the month of April; being then in France; & in the river of Bordeaux; said Joseph met with a severe accident which happened as here after mentioned; the ship had been keeled out for repair, and on righting the ship he was attending to his duty between decks, when he the said Joseph accidently fell into the ships lower hold; and by the fall shattered his thigh badly; he was immediately conveyed to a hospital on shore; wherein he continued with his wound nearly two months; and then returned on board to his duty; although it appeared he was very much injured by the wound he had received from the fall; and still continues disabled by that wound; I must say to do him justice; while he was under my command he behaved himself becoming a faithful soldier to the United States of America & in my humble opinion ought to be disabled with as such by the country; he the said Joseph was in the service of when he received the wound.

Signed Samuel Tucker

Then commander of the Boston Frigate

Author's note – It's interesting that there is no declaration from Joseph Wilkinson detailing his service in his folder, probably because it was burnt in the fire at the War Department. But there was this letter from Captain Samuel Tucker vouching for his disability claim.

Joseph Wilkinson's pension application was approved and he received $8 per month under the 1817 Pension Act.

He passed away the 4th of April 1827.

Jabez Maynard

Author's note – Jabez Maynard's declaration is very long and details land service prior to his sea service. So, I will jump over the beginning of his claim and start at his sea service.

That in the month of February in the year 1779 – engaged as quarter Gunner on board the ship Boston of 36 guns, Capt. Tucker Commander – David Phips 1st Lieutenant, Baxter 2nd Lieutenant – sailed from Boston to Penobscot to guard the coast – took the British frigate Pool of 36 guns from Liverpool – off Delaware Bay and carried her into Philadelphia – sold for $73,000, thence cruised off New York – took the Labinett of 16 guns and a Jamaica Merchantman of 900 hundred Ton. That he was put on board the said prize and was retaken soon after by the British Ship Romolus of 40 guns & was carried into Newfoundland – Lay in prison 8 months & 10 days – was exchanged and returned to Boston – was out about 11 months – repaired ship and went out again with the fleet to Carolina – took a double Decked Brig from Jamaica – was put on board her – and again taken by the ship Romoles and carried into Halifax and lay in prison ten weeks, then exchanged & returned to Boston and joined the ship Dean commanded by Samuel Dickerson Esquire – and went to France in company with the ship Confederacy – on return from France was put on board a prize and was captured by the British Frigate Greyhound – Commanded by Capt. McFarson and carried into St. John's Antigua – Lay in prison three months, was *exchanged & returned to Boston – was out about 15 months – got back about four months after the ship Dean returned. Was slightly wounded several times – was Discharged at Boston the 19th November 1783 by Samuel Dickerson Esquire, commander of the ship Dean – That his Discharge is lost – That he is sixty three years of age.*

Jabez Maynard's pension application was approved and he received $8 per month under the 1818 Pension Act.

He passed away the 1st of January 1841.

USS Cabot

The *Cabot* was a merchant brig purchased by the Continental Congress in 1775. She was 75' in length, had a crew of around 120 men, and was armed with 14 guns.

She is considered by some to be the first armed American vessel to engage the enemy. During an engagement with the British ship *HMS Milford* in 1777, she ran aground and was abandoned. The *Milford* was able to refloat her and she later served in the British navy.[2]

Her captains were said to be J. B. Hopkins, Elisha Hinman, & Joseph Olney

Noah Walrond

Commonwealth of Massachusetts

Dukes County

On this twenty fourth day of October 1832 personally appeared in open court before me George Athearn Esquire Judge of the Court of Probate for Dukes County now sitting Noah Walrond a resident of Tisbury in the County of Dukes County aged eighty seven years who being first duly sworn according to law doth on his oath make the following declaration in order to obtain the benefits of the Act of congress passed June 1832.

That he entered in the revolutionary service in the company of infantry commanded by Capt. Benjamin Smith and stationed for the defense of the sea coast at Martha's Vineyard in said Commonwealth on the first day of June in the year 1776, and served till the first day of September and that on or about the sixth day of September in the same year he entered again the service of the United States as quartermaster aboard the United States Brig Cabot commanded by Elisha Hinman laying in the harbour of Edgartown in said state and sailed on a cruise against the Enemies of the United States and, that we made our course for the English Channel, and that on our cruise took nine prizes, and sent them to the United States and that he came in the prize named the Earlen, and that he continued in the service of the United States with Capt. Elisha Hinman in the Brig Cabot more than two years when he was discharged from the same in New Bedford in the state of Massachusetts.

Author's note – Noah Walrond's pension application was approved and he received $102.60 per year for two years of service under the 1832 Pension Act.

He passed away the 28th of June 1835.

Obed Norton

Commonwealth of Massachusetts

Dukes County

On this 15th day of October 1832 personally appeared in open court before me George Athearn Esquire Judge of the Court of Probate for Dukes County now sitting Obed Norton a resident of Edgartown County aforesaid State of Massachusetts aged 81 years who being first duly sworn according to law doth on his oath make the following declaration in order to obtain the benefits of the Act of congress passed June 1832.

I Obed Norton of the town of Edgartown in the County Of Dukes County do testify and say that I entered volunteer in the service of the State or United States and then resided in Edgartown, County aforesaid as near as my recollection serves me in the month of September 1775 stationed at Holm's Hole in Edgartown, County aforesaid, under Nathan Smith 1st Captain, Benjamin Smith 2nd Captain, Samuel Norton 1st Lieutenant, James Shaw 2nd Lieutenant, and served in the company until about the last of August or the first of September 1776, at which time I obtained leave of the first lieutenant and shipped on board the United Stated Brig Cabbot fitted out in the Delaware and Commanded by Elisha Hinman, then lying in the Harbour of Holme's Hole, Town & County aforesaid, Phillips 1st Lieutenant; do not recollect the names of the other officers, at the time that I shipped on board the Cabbot, Lieutenant Norton made some objection to my leaving the company alleging that the time for which I had enlisted had not expired, to which Captain Hinman replied that his authority superseded Lieutenant Norton's and that if I had a mind to go he would ship me whether the lieutenant liked it or not, said that he was in the Continental Service, that the Navy must be manned and could take men where ever he found them that were willing to go. I then shipped on board the before mentioned Brig Cabbot as Prize Master Pilot for 2-1/2 shares and sailed on a Cruise against the enemy in the then Revolution and served on board of said Brig about four months, at the end of which time I was put onboard a Prize, an English merchant ship as 2nd Pilot, Abner Norton 1st Pilot, the ship being large and unreliable for fear of accident Captain Hinman put two pilots on board and arrived into Providence Rhode Island with said ship on the last of December 1776 or the 1st of January 1777. From which port the squadron all sailed except the Cabbot, and continued on board the Prize Ship until the following spring, I then obtained leave of the agent and returned home, subject to the notice of the aforesaid Agent to return back when called for, but did not get my share of the prize money until the Cabbot & Fleet returned, which was if I recollect right in the Fall of 1777, and was but little the Prize money having been divided amongst the Fleet and we in the Cabbot having made the greatest and most valuable captures. I was paid off all in continental bills. The whole time that I was attached to said Brig being one year or there abouts.

Author's note – Obed Norton's pension application was approved and he received $72.89 per year for his service under the 1832 Pension Act.

There was no mention of his death in his pension application.

Giles Chester

State of Connecticut

County of New London, Town of Groton

On this first day April in the year of our Lord one thousand eight hundred & eighteen before me Elias Perkins Judge of the Court of Common Pleas and for the County of New London personally appeared Giles Chester of said town of Groton & being duly sworn made solemn oath to me that in the year 1775 he shipped as a fifer on board the United States Brig Cabot commanded by Captain John Hopkins & continued to do duty on board said Brig for about 5 months and went in said Brig to New Providence & on his return had an engagement with the British sloop of war Glasgow & came into New London harbor & after refitting went aboard said Brig again under the command of Capt. Elisha Hinman of New London & went on a cruise of about four months & took a number of prizes & returned in said Brig to Boston. The prizes, some of them having gotten into Providence & the deponent further makes the oath that some time in the year 1776 he shipped as a seaman on board the United States Frigate Warren which was built in Rhode Island & commanded by Capt. Hopkins & went out to Bermuda & took a prize, a British ship, that they cruised continued about three months & said Frigate Warren returned to America & arrived in Boston, I was discharged & then shipped a board the United States Frigate Rolla

Author's note – Rolla is corrected on the pension claim in pencil to reflect the correct name – Raleigh.

Commanded by Capt. John Barry & after going out of Boston on the 3rd day out was driven ashore by two 50 gun ships belonging to the British that the deponent was on board the Rolla about four months before he sailed from Boston & was after said Rolla was driven ashore, taken prisoner by the British & carried to New York & put aboard the prison ship called the Jersey, where he continued about five months.

Author's note – Giles Chester's pension application was approved and he received $8.00 per month for 9 months of service on the *Cabot* under the 1818 Pension Act. His widow then reapplied under the 1843 Pension Act and received $36.00 per year for his nine months of service.

He passed away the 10th of February 1838.

Cheney Look

State of New York, Oneida County

On this 18th day of January 1832 personally appeared in open court, before me Nathan Williams, Vice Chancellor of the fifth Circuit of the State of New York, now sitting in chancery, Cheney Look, a resident of Utica in the County of Oneida & State of New York aged about eighty three years, who being first duly sworn according to Law doth on his oath make the following declaration in order to obtain the benefits of the Act of Congress passed June 7th 1832.

That he was born at Tisbury on Martha's Vineyard in the year 1749. That he was aged eighty three years the 17th day of June last. That he has no record of his age or birth. That in the year 1775 – and according to the best of his recollection, in the later part of September or the fore part of October in the same year, he was residing at Edgartown on the Island of Martha's Vineyard, in the now State of Massachusetts, and that he then & there entered the service of the United States, that at that time there were two companies raised, as he believes by order of the government of Massachusetts, which were on the Island of Martha's Vineyard, which where called from their location on the Island the east & middle companies. That the name of the captain of the east company was Benjamin Smith, that the name of the lieutenant was as he believes Malatiah Davis, and the name of the ensign thereof was Samuel Norton. That of the middle company Nathan Smith was Captain & Jeremiah Manter Lieutenant.

That at the forming of the east company at the time above mentioned he enlisted in the same as a volunteer – That these two companies were mustered by Barakiah Bassett who he believes ranked as colonel.

That the company to which he belonged was not engaged in any active service while he was attacked to it.

That he was acquainted with the officers of the above companies, that he has no documentary evidence, & knows of no person whom he can procure to testify to his service in said company, except his brother.

That he left the said company in the month of August 1776. That when he left the said company he furnished in his stead a substitute of the name Crossman, duly armed & equipped. That he received a regular discharge from said company, signed by Benjamin Smith, the captain hereof, which said discharge he is now unable to find, but by what means it has been lost he does not know.

That the reason of his leaving the said company was as follows – In the month of August 1776 the Brig Cabot of New London – a vessel of war, belonging to the United States, mounting fourteen guns, came to Martha's Vineyard and anchored in the harbour of Edgartown where the company to which he belonged was stationed. That the said ship was not well manned and the captain thereof was desirous of enlisting recruits – That this declarant wishing to engage in more active service than the troops at Martha's Vineyard were engaged in procured his discharge from the company to which he was attached in manner aforesaid, and immediately enlisted as a common seaman on board said Brig and served accordingly – That the Brig sailed from Martha's Vineyard in August 1776 for the purpose of intercepting the Jamacia Fleet, being British Merchantmen homeward bound – That after about fourteen days sail the Cabot overtook said Fleet – That the Cabot in the course of the cruise took five vessels of the said Fleet and sent four of them home as prizes, that this deponent was sent home with the Barque Lawthar, one of the prizes, which was commanded by Jonas Hamblin, the fourth Lieutenant of the Cabot. That the Lawthar put in at Providence about the first of November 1776 when this declarant left the service, having shipped for that cruise and did not enter the service again during the war.

Author's note – Cheney Look's pension application was approved and he received $48.00 per year for 11 months as a private & 3 months as a seaman under the 1832 Pension Act.

There is no mention as to his date of death.

Caleb Prouty

State of Massachusetts, County of Plymouth

On this twenty seventh day of August A.D. 1832 personally appeared in open court before the Hon. Willis Wood Judge of the Court of Probate for said County of Plymouth now sitting at Hanover in said county, Caleb Prouty, a resident of Hanover situated in the County of Plymouth and State of Massachusetts aged eighty five years, who being first duly sworn according to law doth on his oath make the following declaration in order to obtain the benefit of the Act Of Congress passed June 7th 1832.

That he entered the service of the United States under the following named officers and served as herein stated.

About the middle of April 1775 I enlisted in a company commanded by Samuel Stockbridge and was stationed upon the beach between Scituate Harbor and the Mouth of the North River so called dividing Scituate and Marshfield to watch the movements of Capt. Belfour who had landed at Marshfield with a company of British Tories called the Queens Guards. I there served as a private soldier two weeks, at which time said Belfour with his company embarked on board an English ship in the bay and I was discharged.

Again in 1775 I enlisted in Capt. Joseph Soper's company and was stationed at Scituate Harbor and on the beach as above stationed and served as a private volunteer three months guarding the harbor and was then discharged by Capt. Soper, nothing particular occurred and other officers not recollected.

Again I think in the year 1776, but of what year I am not positive, I enlisted on board of a Brig called Cabbot, but whether said Brig belonged to the state of Massachusetts or was a continental vessel is not within my recollection. The Brig was commanded by Capt. Joseph Olney, or however otherwise his name may have been spelled, and served as cook for him of seven months. The Brig lay in Boston harbor, after I enlisted about six months during the winter and went to sea about the first of March, but soon fell in with a British frigate called the Milford and driven on shore in the Bay of Fundy, near Cape Sable. The Brig was afterwards got off by the Milford, and myself with the rest returned to Boston, where I was discharged.

Author's note – Caleb Prouty's pension application was approved and he received $4.42 per year for 3 months and 15 days as a cook under the 1832 Pension Act.

There is no mention as to his date of death.

USS Columbus

The *Columbus* was a private ship purchased by the Continental Congress in 1775. She was manned by a crew of around 220 men and armed with 18 guns.

In 1778 she was run aground and burned by the British.

Her captain was said to be Abraham Whipple.

Columbus – U.S Navy & connecticuthistory.org

Henry Malcolm

District of Pennsylvania

On this seventeenth day of April A.D. 1818 before me Richard Peters Judge of the District Court of the United States for the Pennsylvania District personally appeared Henry Malcolm of the city of Philadelphia, who being duly sworn depose and declares that during the war of the revolution he served against the common enemy as a Surgeon in the naval service of the United States – That on the 28th day of December AD 1775 he received a commission as Surgeon on board the Ship Columbus of which Abraham Whipple Esquire was commander, by the proper authority delegated by the Continental Congress for that purpose – That he served with this Ship in the Squadron under the command of Commodore Hopkins employed in the raid of the British Island of New Providence in the year 1776 – That on his return from the raid expedition, he was ordered to fill the station of Surgeon on board the United States Ship Andrew Doria, Captain Biddle, and in that capacity performed a cruise of two months and more – That after his return to his country in the fall of the year 1776 he married in the State of Rhode Island and he there upon had permission to retire from the naval service of the United States, after having faithfully served for the term of nine months and longer.

Author's note – Henry Malcolm's pension application was approved and he received $20.00 per month for 4 months and 18 days as a Surgeon under the 1818 Pension Act.

There is no mention as to his date of death.

Coggeshell Butts

Declaration

In order to obtain the benefit of the Act of Congress passed June 7th 1832.

State of Rhode Island & Providence Plantations, County and town of Bristol.

On this first day of October A.D. 1832 personally appeared in open Court before the Judges of the Court of Probate for the town aforesaid, now sitting at Bristol aforesaid, Coggeshall Butts, a resident of said town & County of Bristol, aged Eighty One years, who being first duly sworn according to law, doth on his oath make the following declaration in order to obtain the benefit of the Act of Congress Passed June 7th 1832. That he entered the service of the United States under the following named officers and served as herein stated, that he was born in Portsmouth on the island of Rhode Island on the 27th day of April 1751 – That at the commencement of the Revolutionary

War he was residing in and sailing from the town of New port, that in the month of October 1775 as near as he can recollect, Commodore Hopkins wrote to Newport that he was in want of fifty men for a naval expedition, then preparing at or near Philadelphia, Commodore Hopkins then being at Providence, upon this information he the said Butts, went immediately to Providence as a midshipmen on board the American vessel of war called Columbus under the command of Abraham Whipple, which ship with the Alfred and several other smaller armed vessels made up the squadron commanded by Commodore Hopkins – That he was to receive twelve dollars per month during the expedition – That he repaired immediately for the squadron, them at Philadelphia and afterwards departed down to and sailed from a place called Reedy Island in the Delaware River – After all things had been got ready which did not take place until very cold weather, nor until we had been frozen in for a considerable time, which was until sometime in February 1776, in the course of which month as near as he can recollect the squadron sailed for New Providence, where they landed and captured a number of cannon, and other articles of war, which they brought away together with the governor & several other public officers, all which were handed over at New London – That said squadron also captured on their way home a British Brig & schooner, the first of which was a vessel of war, they also had an engagement with an English Ship of War off Block Island, in which the squadron lost several men and among there wounded was a son of the Commodore, that the squadron remained in New London a number of weeks, when it was ordered to Providence to refit and he the said Butts continued on board and in the service of the United States until late in July 1776, when he was discharged, but he has no recollection of taking any written discharge. One Rhodes & one Arnold were lieutenants on board the Columbus, the Boatswain's name was Baker; Cornell Gill & Granston were midshipmen with himself. The English ship we engaged was called the Glasgow.

Author's note – Coggeshall Butt's goes on to explain a brief tour of duty as a private on land.

His pension application was approved and he received $65.00 per year for 1 month and 45 days as a private and 10 months as a midshipman under the 1832 Pension Act.

There is no mention as to his date of death.

USS Confederacy

The *Confederacy* was a 36-gun frigate purchased by the Continental Congress in 1778. She sailed under an American flag unit she was captured by the British in 1781.

Her captain was said to be Seth Harding.

Confederacy – Connecticuthistory.org

Thomas Edgar

State of Connecticut, New London

Be it remembered that on this 24th day of March 1818 before me Jeremiah G Brainard a Judge of the Superior Court of the State of Connecticut personally came & appeared Thomas Edgar of New London in said State & being duly & solemnly sworn according to law and the usage of said State did on his said oath by me to him administered, deposed, & say that he shipped as carpenter on board the Continental Frigate Trumbull, James Nicolson Esq commander on or about the 15th day of September 1779. That he sailed in said frigate as carpenter for up wards of one year, then after during which time he was engaged with the British Letter of Marque Watt. That when the Trumbull was laid up at Philadelphia he was transferred to the Continental ship Confederacy, Capt. Seth Harding. & in her was captured by the British ship Roebuck of 50 guns & another the name of which is forgot, was sent to New York confined on board the Jersey Prison Ship, was taken thence & sent to England in the Europe of 64 guns. Confined in Mill prison and afterwards exchanged & landed at Boston in August 1782.

Author's note – Thomas Edgar's pension application was approved and he received $8.00 per month under the 1818 Pension Act.

There is no mention as to his date of death.

John Shober

Author's note – I would normally skip over lengthy land service at the beginning of a pension claim, but Mr. Shober's is very interesting and I have included it.

State of Virginia, County of Berkely

On this 8th day of July 1833 Personally appeared in open Court being a Court of Record John Shover (at present spelled Shober according to the German) a resident in said county aged 73 years the 17 day of August last past, who being first duly sworn according to law doth on his oath make the following declaration in order to obtain the benefit of the Act of Congress passed June 7th 1832.

That he the said John Shober at the time the British over ran the Jerseys volunteered his services & served three months, which was in the winter of 1776 & 1777. That he was in the skirmish at Burbble Town where three of his mess were killed – That in the month of September following at the time the British landed at the Head of Elk he again volunteered his services & marched in the company commanded by Capt. Hoff, Baker Johnston Colonel of the Regiment Maryland Militia, and under the command of General Smallwood. That the division arrived at Read Quarters on the Bergomy, and that on the night of the 3rd of October the army took up the line of march & at day break arrived at the foot of Chesnut Hill, where he was one of 40 men picked out by the Captain to commence an attack on the out post of the enemy. The morning being very foggy we advanced too far & received the enemy's fire on our left flank within a few yards. The Captain was wounded, Lieutenant Gosh fell by my side & myself received a ball through my thigh. Although wounded I advanced on the enemy & discharged my piece 5 times, being irritated by the pain of my wound. During the engagement four balls more entered my clothes & grazed my body without inflicting another wound accepting a piece of melted lead or slug entered under my rib where it remained for years before it was extracted. That he was left on the field when the army retreated, being chased by the British Horse. He took refuge on a rock and kept them at bay, he left the rock and roads & passed through the woods, he arrived the third day at Head Quarters, his wound being irritated & inflamed for want of dressing. He was put in the Hospital at Reading where he remained some time when he obtained a passage home in a wagon. & for six months he was disabled from work on account of his wound having been improperly dressed & the fracture of the nerves & on the wound.

And further that he the said John Shober in the latter part of 1780 enlisted on board the Confederacy Frigate of 36 guns then lying at anchor opposite Chester in Delaware – (being enveighled on board & compelled to enlist) That he served in said vessel until the arrival of the same at Cape Francis in the island of Hispaniola, that he was attacked by the prevelant fever & that he together with some of the crew were landed & placed in the Hospital – Meanwhile he was informed by some of the French officers of the Navy (having seen none of the officers belonging to the Confederacy for nearly a month), that a British Frigate hove in sight off the Harbour & that the Confederacy went out to meet the British vessel & that it was a general report & fully believed that the Confederacy struck her flag without making a defense "The French officers further remarked that as soon as I was fit for service I would be put on board of a man of war (saying we fight for the american & the americans must fight for us) Not willing to serve the French & hearing nothing further of the Frigate, the officers & crew, & there being no American vessel in the Harbour, he left the Hospital & travelled through the island in a weak & forlorn state, passing though Ganives, Port Lemar & Semaar & other ports seeking for american vessels, that some time in the month of April or May 1781 he arrived at Port au Prince, where he joined the BrigMars of 12 guns commanded by Capt. Gannon, bound for Philadelphia – He agreed to work his passage & on those terms he entered the Brig, Capt. Gannon also informed him that the Confederacy was captured as aforesaid. Leaving Port au Prince we were chased by British cruisers so that we were compelled to seek safety in the island of Curaco, having touched at St. Eustatius, but the British fleet under Heelney had taken the island & we narrowly escaped being taken.

Arrived at Curraco & I remained until September when an American Brig named Two Friends sailed for Philadelphia on which I entered by working my passage & that sometime in the latter part of October 1781 I landed at Philadelphia being absent nearly between 10 & 11 months. I made enquiry respecting the Frigate & was informed as afore stated. He was advised to call on Mr. Morris for his Bounty & pay (of which he had received no part) Mr. Morris informed me that he was ruined & could not do any thing for me. He then returned to Frederick Town state of Maryland the place of his nativity, from whence he removed in 1785 to Martinsburg, Berkeley County Virginia where he at the present resides.

To corroborate the foregoing statement & throw as much light on the subject of the truth of the case & of his actual services, he will state a few facts from memory, he thinks the name of the Capt. of the ship & Capt. Of the marines were Harding or Hardin. The armourers name was Ben Rothrock, one of the Boatswains was called Portugee John & the cook being remarkable, having lost an arm –

The Confederacy, Fair American & Saratoga Frigates left the Capes of Delaware together & for some time sailed in company, after passing Turks Island & being in the Gulf of Florida & being overtaken by a storm the fleet was scattered & the two latter vessels were lost as they were not heard of since.

Author's note – John Shober's pension application was approved and he received $53.33 per year under the 1832 Pension Act.

He passed away the 3rd of October 1836.

Henry Norris

State of Connecticut, New London

On this 26th day of March 1818 before me Jeremiah G. Brainard a Judge of the Superior Court of the State of Connecticut personally came Henry Norris of the town & county of New London in said state & being duly sworn according to law and the usage of said state due on his oath by me to him administered declose, depose & say that he on board the Continental Frigate Trumbull, James Nicholson Esq commander about September or October 1779 in said frigate as Wardroom Boy, was in her when she was engaged with the British Letter of Marque ship Watt, returned in her to Boston where she under went some repairs & then went in her to Philadelphia where he was discharged after having served on eight months & immediately at Philadelphia shipped on board the Continental Ship Confederacy, Seth Harding Esq commander & sailed in her as cabin boy on her cruise & off Hispaniola was captured in her when homeward bound by the British Ship Roebuck & the British ship the name of which he does not distinctly remember & was after wards confined on board the Jersey Prison Ship in New York for more than six months when he was exchanged, having served in the whole in a continental vessel nearly or quite two years during the war of the Revolution.

Author's note – John Shober's pension application was approved and he received $8.00 per month under the 1818 Pension Act.

There is no mention of his death.

Cornelius Wells

District of Columbia, Count of Washington

On this 15th day of December 1832 personally appeared in open court before Judge of the circuit court of Washington, now sitting, Cornelius Wells, a resident of the District and County aforesaid aged Eighty years, who being first duly sworn according to Law, doth on his oath make the following declaration in order to obtain the benefits of the Act of Congress passed June 7th 1832.

Author's note – Mr. Well's starts off with lengthy land service, which I will omit and jump into his naval service.

On to the fall of 1779 when I was transferred to the navy, on board of the United States armed ship the Confederacy, commanded by Capt. Seth Harding. In consequence of the death of the master at arms – I was appointed master at arms eleven days after my going on board of said ship. On the first cruise lost a mast and put into Martinique to repair – Then *returned to Philadelphia, at the Capes of Delaware we sent on shore a pilot, in attempting to land, the boat was capsized and the 1st Lieut. Wan and some of the men were drowned. On our second cruise captured several vessels among them one with 365 slaves and some gold dust. Took her into Cape Francis and there sold ship & cargo. Then took in continental stores and sailed for Philadelphia, on our passage was captured by the Robuck of 44 guns and the Alfus of 38 guns – of the Royal Navy. They carried us into New York where I lay in prison on board of the ship Jersey the winter of 1780 and 1781. In the spring of 1781 we were distributed on board of the transport vessels. Was then put on board of his Majesties ship Robuck of 44 guns, Capt. Cosby commanding. In Nov 1781 the British fleet returned to New York when all that would not inlist in his Majesties service were sent back to prison, of which number I was one. Lay in prison until the spring of 1783 when I was exchanged and sent to Elizabethtown& from there to Philadelphia. Was then put under the command of Capt. Barney, the ship Hyde Alley mounting 23 guns in the spring of 1783. Peace being proclaimed I reported my self to the Board of War, settled my account, and received my pay as master at arms at the rate of $18 per month in continental money.*

Author's note – Cornelius Well's pension application was approved and he received $120.00 per year under the 1832 Pension Act.

He passed away "on or about" the 21st of July 1844

USS Dean & Hague

The *Dean* was an armed merchant frigate loaned to the Continental Navy by the French in 1778 and armed with 34 guns.

In 1782 she was renamed the *Hague* and in 1783 she was taken out of commission.

Her captain was said to be Samuel Nicholson & John Manley.

You are going to read in some of these claims that Samuel Nicholson was relieved of his command as captain of Dean. This is true. He was accused of mistreatment and poor leadership by one of his officers. He was relieved of command, went to court martial, and was found guilty. The court martial was later appealed and he was acquitted. He went on to captain the *USS Constitution*.[3]

As for the name change, there is speculation that after the relief of Samuel Nicholson from *Dean*, the ship changed its name to *Hague* to shed its bad reputation.[2]

Peter Wakefield

State of Ohio, Lawrence County;

Court of Common Pleas of August 1832

On this 28 day of August 1832, personally appeared before the judge of the court of common pleas, now sitting in & for the county of Lawrence, Peter Wakefield, a resident of Windsor Township in the county of Lawrence, aged 68 years – who being first duly according to law – doth on his oath make the following declaration in order to obtain the benefits of the provision made by the Act of Congress passed June 7, 1832 –

Author's note – I am going to skip his first land service and jump right into his sea service.

In Boston Massachusetts – about the 10 August 1782, he again enlisted for the term of five months, to serve on board of the Continental Frigate – then called "Dean, afterward called the Hague" – 36 guns – commanded by Captain Nixon he thinks, who shortly after was arrested - & afterwards the ship was commanded by Commodore Manley -Sailed from Boston Harbour went to the West Indies – cruised till after Peace, till the 8th of April 1783, when news of the Peace was received.

The ship took five prizes – one Letter of Marque by the name of "Baily" of 20 guns – the others were Merchantman – After they took the Baily, they were chased by a 50 gun ship two days

– were driven into Bay Maho, Guadeloupe, when the Hauge run aground – The British ship came about half a mile – could get no nearer – Sought us three days – attempted twice to board the Hague & men repulsed – they then lightened the ship, and by help of boats from the island got out of the reach of the enemy & escaped – 4 other British ships gave chase before the Hague got into the Bay – they also lay near us & fired upon us for three days – They attempted twice to board with 15 boats at a time – which were beaten off with great loss – The ships opposed to us in this action were 1,500 guns – 1 at 64 – 2 at 74 & 1 at 84 guns – name of the ships unknown – The officers of the Hague were Commodore Manley -1st Lieut. Page – 2nd Lieut. Waterman – Declarant served as a marine – From Guadeloupe sailed to Martinique – Fort Rial – then repaired – I went on another cruise on the same coast – on the news of Peace, returned to Martinique and took in the sick & wounded, theft then & on the 8th April sailed to Boston – arrived the 2 May – was discharged about the middle of May 1783.

Author's note – Peter Wakefield's pension application was approved and he received $41.66 per year for 12 months & 15 days of service under the 1832 Pension Act.

There is no mention as to his date of death

Jonathan Wilkins

I Jonathan Wilkins a resident citizen in the town of Mount Vernon in the County of Hillsborough in the State of New Hampshire, Pensioner, sixty nine years of age the 16th of September last past do solemnly declare that in the year of our Lord 1775 I turned out immediately after the Lexington Battle as a volunteer in the service to my country in the latter part of April of that year I enlisted for the term of eight months into the company commanded by Capt. Archelans Towne in the Regiment commanded by Col. Eben Bridges at which time I faithfully served out & was honorably discharged by my officers.

In the month of April in the year of our Lord 1782 I enlisted & shipped on board the Continental Frigate Hague commanded by Capt. John Manley. On this cruise I was out sixteen months. During the cruise I was wounded at the time Admiral Rodney's Fleet attacked our frigate, which to the best of my recollection was on the ninth or the eleventh day of June AD 1783 in a place called Bay Mayo, or some place of a similar name. In the best of my recollection we escaped by sailing over a sand bar, tho' we were exposed to the fire of the enemy.

I further declare that I am placed on the pension list of invalids of my county at a rate of two dollars & fifty cents per month, which has been lately raised to four dollars per month, which pension I shall relinquish should I be placed on the pension list under the late act of Congress.

In June or July 1783 to the best of my recollection we landed at Hancock's Warf in Boston Massachusetts, the place we started from, where I was honorably discharged by Capt. Manley's clerk, one Richard Con, which discharge I have lost.

I further state that I need assistance from my country for my support being unable to work and am now dependent & helped by the charity of the town.

Author's note – Moses Ayer's pension application was approved and he received $8.00 per month for one year of service as a marine under the 1818 Pension Act.

He passed away the 27th of February 1823.

Peter Powers

State of Vermont, Rutland County

On this third day of January 1820 before me the subscriber one of the assistant Judges of the County Court for the County of Rutland is said state, personally appears Peter Powers aged fifty four years resident in Pittsford in the county & state aforesaid, who being by me first duly sworn according to law doth on his oath make the following declaration in order to obtain the provision made by the late Act of Congress entitled "An act to provide for certain persons engaged in the land and naval service in the United States in the Revolutionary War."

That said Peter Powers enlisted in the month of May or June 1782 in Hollis in the State of New Hampshire as a marine on board the Frigate Hague, formerly the Dean, commanded by John Manley. That he entered on board said frigate in June or July 1782 and that he continued to serve on board said vessel until May 1783 when he was discharged from said service in Boston, State of Massachusetts. That he was in the Battle at the taking of the Baliff, as he understood she was called, a 20 gun ship and sent her into Boston. Also 4 other prizes, a Brig & sloop taken off against Antigua and the next day a Brig & another Brig or sloop. And that in January or February we espied a sail in the morning. She proved to be a fifty gun frigate and gave chase. She chased us until the next morning where she was joined by 2 seventy fours. We lightened our frigate by throwing our guns overboard except 8 or 10, which enabled us to run over the bar into Point Petre and grounded soon after crossing it. The seventy fours made off but soon returned with two others about the same heft, which made five in all that played upon us off & on 48 hours or more, sometimes with their heavy guns, at other times they attempted to board us with their boats. Finally we had the good fortune to get out of the mud and arrived safely in town.

Author's note – Peter Power's pension application was approved and he received $8.00 per month for one year of service as a marine under the 1818 Pension Act.

He passed away the 11th of April 1854.

Luther Dana

Massachusetts District

Comes – Luther Dana – now resident of Cambridge in the County of Middlesex Commonwealth Massachusetts Esquire – He declare and say that in the month of May one thousand seven hundred and eighty two (to best of his recollection) he entered as an Acting Midshipman for the United States Frigate Dean – and was employed in recruiting for said ship by order of her commander – Samuel Nicholson Esq. till about the middle of the month of June – when he joined the frigate aforesaid in the Harbour of Boston, John Page Esq. being her First Lieutenant – That he was attached to and was on board said Frigate doing duty as an Acting Midshipman Volunteer – or Midshipman till close of the war – and after – Except when detached from her on the recruiting service or other special duties by order of the commander – That previous to her sailing on a cruise from the Port of Boston – Samuel Nicholson Esq. was suspended from the command of said Frigate – John

Manley Esq. was appointed and took the command and the Frigate name was changed – and she was called and know, after, by the name of Hague, under the command of said Capt. Manley. We proceeded in said Frigate on a cruise against a common enemy, the British – and captured several of their vessels which became Prizes – This deponent saith – That he was ordered on board (by the commander) of the second Prize as Chief Mate – Daniels a volunteer & Prize Master on board said Frigate Hague have charge of said capture – We were ordered to proceed for and arrived at St. Pierre Martinique when she was condemned as lawful Prize. That after the said Frigate Hague arrived at St. Pierre – that this deponent with others in said Prize were by Capt. Manley ordered to repair on board said Frigate and join her in our stations. That on joining said Frigate the commander Capt. Manley did present this deponent with a document making him a Midshipman in the Navy of the United States in the month of December 1782 – That soon after he proceeded in said Frigate as a Midshipman on a cruise under the command of said Capt. Manley and they fell in with and captured the Bailley of London, a Letter of Marque mounting twenty guns on one deck with a full compliment of men – That this deponent was ordered on board to take charge of her as Second Officer, the aforesaid Daniels being Prize master – and they were ordered for and arrived at the Port of Boston – where said capture was condemned as Prize – Thomas Russel Esq. being the U. States agent – That this deponent was directed by him to remain in charge of said prize until sold, which he did – there being no vessel of the Navy of the U. States in said Boston

The Frigate Hague – which this deponent belonged to returned to Boston in the month of May one thousand seven hundred and eighty three – when this deponent received an order from the commander to repair on board- which he did – doing duty as a Midshipman till legally discharged – when said Frigate was sold out of service – the Navy of the U. States being dismantled.

Author's note – Luther Dana's pension application was approved and he received $8.00 per month for one year of service as a midshipman under the 1818 Pension Act.

He passed away the 19th of February 1834.

Oliver Holden

Commonwealth of Massachusetts, County of Middlesex

On this sixteenth day of February in the year of our Lord eighteen hundred and thirty six personally appeared in open Court, before the Honorable Judges of the Court of Probate now sitting at Charlestown in said county a resident of Charlestown in the County of Middlesex and Commonwealth of Massachusetts, aged seventy years, who being first duly sworn according to law doth on his oath make the following declaration, in order to obtain the benefit of the act of Congress, passed June 7th 1832.

That he entered the service of the United States under the following named officers and served as herein stated.

That is to say, In the early part of the spring of A.D. 1782 he was living with his father in Pepperell. Mass and Lieutenant Reed came up here as a recruiting officer and he enlisted under him as a Marine on board the Dean Frigate for the term of one year, and received from him a bounty of ten dollars.

Author's note – a "bounty" was a monetary signing bonus for joining.

Soon after this he joined the Frigate in Boston Harbour under the command of Capt. Nicholson, who was afterwards tried by a Court – Martial and Capt. Manley then took command of her, and her name was changed to the "Hague." She sailed in the month of August on a cruise among the West India Islands. They captured four British prizes and made the Port of St. Pierce Martinique. The morning after they sailed from this Port, after a short conflict they captured a British Letter of Marque, called the Zac. Bailly of twenty guns. He was placed on board this ship of which Capt. Daniels was Prize Master and Audebit and Dana Mates. They returned to Boston where they arrived in the latter part of the winter of 1783. They remained on board for some time & the vessel was discharged, and he received his discharge from the hands of George Richards who had returned in the Frigate. Richards was Chaplain, and acted as Assistant Captains Clerk. He thinks this was sometime in April 1783. And after he had been in service about one year – Palms was Captain of his company and Reed and Waterman his lieutenants – His brother Richard Holden was Orderly Sergeant. He recollected Cochran was boatswain and that one McKeever was Sailing Master.

Author's note – Oliver Holden's pension application was approved and he received $110.00 per year for one year of service as a marine under the 1832 Pension Act.

There is no mention of his date of death.

USS Fly

The *Fly* was a former merchant sloop purchased by the Continental Navy in 1775. She carried 6 cannons and sailed under an American flag until she was burnt to prevent being captured by the British in 1777.

Her captains were said to be Hoysted Hacker & Seth Harding.

The following is a painting of the USS Fly & USS Mosquito, which you will read about later, by William Nowland Van Powell

Aaron Warren

State of Maine, County of York

On this third day of September 1832 personally appeared in open court, before the Court of Probate, now sitting at Alfred, Aaron Warren, a resident of Wells in the County of York and State of Maine, Physician, aged seventy four years, who being first duly sworn according to law, doth on his oath, make the following declaration, in order to obtain the benefit of the Act of Congress passed June 7, 1832.

Author's note – Mr. Warren's pension declaration is very long. So long that I am going to jump to page 5 of his declaration when he gets into his service aboard *Fly*.

There was also in the service of the United States in year 1779 as follows. In the summer of that year he sailed in a private vessel from Boston on a cruise and ran ashore and lost their vessel, a little south of Newport – From thence he went to Philadelphia in order to get into Business again. He was applied to by Captain Robinson in the service of the United States to go with him in the sloop Fly of four guns, which was a tender to a United States Ship called, he believes, the Confederacy, then lying in the Delaware River – She was commanded by Captain Hardin. He was applied to, to enter on board said vessel as Surgeon. Captain Robinson requested him to go to and be examined. He went to George Glentwortt, Surgeon of the General Hospital and received a certificate of which said the following "Philadelphia Sept. 25 1779. At the request of Capt. Robinson I have examined Doct. Aaron Warren on the subject of Physic and surgery and find him duly qualified to act as Surgeon of the Fly Sloop. G. Glentwortt S.S.G.H. Navy Board." I immediately went on board, having no other commission. There was a Lieutenant on board by the name of Robinson – there was no other officer. – there were about thirty men. They cruised up & down the river while the ship lay at anchor – When at the mouth they were blown off. The captain during the gale told the Pilot, whose name was Goforth, to stand off so many hours and then call him, he then retiring to his berth. The Lieutenant having got more than half of the crew with him, being an Englishman who had engaged in our service, took over the vessel before the captain was called and put her away for St. Augustine – They were five days and nights scudding in a gale of wind- When near St. Mary's they fell in with an English schooner bound to Newport, commanded by Captain Millray, bound to Newport. Her mainmast were gone, having been carried away in the same gale. The Captain, Pilot and himself obtained permission to go on board – They went in this schooner to Tibey Light House where there was a French fleet laying.

Author's note – I think he is referring to Tybee Light House in Georgia

The Captain supposing it to be an English fleet, went in and was taken. The fleet was commanded by Count Duryman and was part of the fleet under the command of De Estang – They were just weighing anchor to go to Charleston S.C. We were

taken aboard the Counts ship and arrived off Charleston in four or five days – When the said Warren, Robinson, and Goforth went to John McCalla, a Notary of the Public and entered their protest against the mutineer.

Author's note – Mr. Warren goes on to detail more sea service, which I have omitted.

His pension application was approved and he received $312.39 for his service under the 1832 Pension Act.

There is now mention of his date of death.

Joseph Breed

State of Pennsylvania, Venango County

On this twenty ninth day of August in the year of our Lord one thousand eight hundred and thirty two, personally appeared in open court before Henry Shippen, Esquire, President and John Irwin and Thomas McKee Esquire, his associate Judges of the Court of Common Pleas of said county now sitting Joseph Breed, a resident of Cherry Tree township in the said county of Venango and State of Pennsylvania aforesaid, aged seventy four years, who being first duly sworn according to Law, doth on his oath make the following declaration in order to obtain the benefit of the Act of Congress passed June 7th 1832.

That he entered the service of the United States under the following named officers and served as herein after stated – That he enlisted in the Rhode Island in the month of September 1775 and entered in that service for about two months under Lieutenant Poly Berry in Captain Thomas Wells company, and then some time in the month of November in the same year enlisted in the continental sea service under William Grinnell who was recruiting officer. Went on board the Sloop Fly commanded by Hoisted Hacker. The fleet then lying about thirty miles below Philadelphia, and under the command of Commodore Erick Hopkins. We then sailed for New Providence and captured it, still under the command of Commodore Erick Hopkins – Then returned some time in the month of March to New London (in Connecticut) with the spoils we had taken at New Providence.

Author's note – Mr. Breed goes on to detail more land & sea service, which I have omitted.

His pension application was approved and he received $71.66 for his service under the 1832 Pension Act.

He passed away the 23rd of January 1839.

USS General Washington

I could find no information about the *General Washington*, but it is listed on several sources as having existed. And by evidence of these pension claims, it was in service with the Continental Navy.

Her captains were said to be Joshua Barney, Samuel Walker, Jonathan Caldwell & Thomas Hardy.

John Manley

To the Honorable Senate and House of Representatives in Congress assembled

The Petition of John Manley sayeth that your Petition during the Revolutionary War was a Midshipman in our Navy and under the Command of Commodore James Nicholson of the Governor Trumbull Frigate, then in the port of Philadelphia. He entered the service the winter of Eighty and eighty one, then undergoing heavy repairs from damages received in her action with the Watt on the 2nd of June previous, consequently we was not prepared for the sea until late in the spring, and when passage was blockaded by the enemy, the Media &

Amphratrite, two frigates – Sail'd from Cape May on the tenth day of August with several armed vessels, among the numbers the Royal Lew's of 20 guns, Steven Decatur , the father of the "late hero."

Author's note – He references the *Royal Louis*, which was commanded by Steven Decatur Sr. His comment regarding the "father of the late hero" is in reference to Steven Decatur Jr., his son, who was a naval hero in the War of 1812.

On our clearing the Cape we saw three vessels of war to the Eastward. They made sail after us, we making our offering with a brisk gale from the westward. At 3pm we was dismasted in the chase, fore top mast went over the side, by the fore cap our head sails all went with our main top gallant sail with them. We could sail no other way than before the wind. At midnight they came alongside, the frigate hailed us, upon answering they gave us a broad side, we battled three glasses,

Author's note – In order to tell time at sea, an hour glass filled with sand was used. An hour was referred to as a "glass."

Was compelled to strike, our captures proved to be the Iris Frigate & Gen. Monk sloop war, Dawson commander of the first, John Rogers the latter. We was roughly handled, 10 or 12 killed, and as many wounded. We were towed into New York. The crew disposed off in the Prison Ships, the Old Jersey principally. Officers on parole on Long Island. It was my lot to spend between three or four months on board the Old Jersey, our officers except three that belonged to the Eastward returned to Philadelphia, their being no public vessels Mr.

Robert Morris our Navy agent propose to us to make a short voyage to Havanna. We did. Richard Dale 2nd Lieutenant, Capt. Murry volunteered, and several others, myself amongst the number. On my return I was attached to the sloop of War General Washington, Joshua Barney Esq. Commander, bound to L'Orient with dispatches for our ministers then at Versailles. And with him continued running to Hasre & Plymouth until the 4th day of April 1784. Then the ship's crew was discharged, ship sold by order of Robert Morris. We then laying at Baltimore. This was the last ship in commission of the American Navy. I kept a regular journal while on board this ship of various transactions and look much plans to keep it in good order from that time until a few years since a few years since, but owing to several years of neglect and carelessness of my friends, they was mutilated is such a manner that I have but eight days reckoning entire.

Author's note – There were several declarations in Mr. Manley's folder and the above was the best written. But on another declaration, he further states the following about his tour on *General Washington.*

In the fall of 1882 was attached to the United States ship of war Genl. Washington, Capt. Joshua Barney ESG. Commander; Said ship was employed as a packet to Europe with dispatches and continued until April one thousand seven hundred and eighty four; She was the ship that brought over the preliminaries of peace from L'Orient and also a heavy loan of money, which your petitioner deposited in the Bank of North America, Chestnut Street, by order of Robert Norris Esq. our financial agent. This ship made three voyages to Europe, the first to L'Orient, the second and third to Hasre and Plymouth, and further before we departed from L'Orient there was forwarded to Capt. Barney through the hands of Benjamin Franklin from Versailles a passport from the "Crown of Great Britian" for our ship having "dispatches of high importance" and proceed home immediately.

Author's note – John Manley's pension application was approved and he received $144.00 per year as a midshipman under the 1832 Pension Act.

There is no mention as to his date of death

Benjamin Berry

Commonwealth of Massachusetts, County of Barnstable

On this twenty ninth day of October in the year of our Lord one thousand eight hundred and thirty three personally appeared in open court before the Honorable Judge of the Court of Probate now sitting at Brewster in and for said County, Benjamin Berry, a resident of Brewster in said County, aged Sixty eight years, who being first duly sworn, according to law, doth on his oath make the following declaration in order to obtain the benefit of the act of Congress passed June 7th, A.D.1832.

That he was born in Harwich, now Brewster, in the County aforesaid on the twenty sixth day of July A.D. 1766.

That he was in the United States Service from the 31st May 1779 to the 31st Aug 1779 as a seaman on board Brig Active, Allen Hallet, commander. He shipped on board said Brig at Boston in May aforesaid, she being then a private armed vessel. He said from thence to Portsmouth for repairs, where laying there she was ordered to join the expedition at Penobscot in the Government service as he thinks that they sailed from thence to Townsend in the District of Maine, that he served there with other vessels until the reinforcement of the British came & we were driven up the River where the fleet was destroyed & we were discharged & travelled to Boston and thence home.

Also, that he was in the United States service (if the ship hereinafter mentioned was a stated ship as the declarant believes was the fact) from the 1st of March 1781 to about the last of September 1781. He enlisted on board a ship of twenty guns called the General Washington or George Washington, Samuel Walker, commander. He enlisted at Harwich aforesaid & proceeded to New Bedford where he entered on board said ship & went to sea in her about the 1st of May & after cruising & taking one Brig, we were captured by the ship Chatham & carried into N. York & put on board the prison Ship Jersey. On the 15th of Sept. the sick were exchanged of which member I was one. We were then brought to New London & then got home on foot as well as we could & arrived home about the last of September 1781.

Author's note – Benjamin Berry's pension application was rejected because he did not prove he served for more than 6 months.

He appears to have passed away on the 15th of September 1842.

Daniel Hilyard

State of New York, County of Clinton

On this fourth day of October in the year of our Lord one thousand eight hundred and thirty two personally appeared in open court before John Palmer, John Warford, Seth Calkins and Miles Stevenson and James W Wood Esquires Judges of the Court of Common Pleas of the County of Clinton now sitting, Daniel Hilyard, a resident of Plattsburg in the said County aged seventy three years, who being first duly sworn according to law doth on his oath make the following declaration in order to obtain the benefit of the act of Congress passed June 7th 1832.

Author's note – I am going to jump over Mr. Hilyard's initial land service and join his declaration in January 1778.

He again enlisted at New London to go on board the Oliver Cromwell, a twenty gun ship belonging to the State of Connecticut, commanded by Capt. Timothy Parker, 1st Lt. Chapman, 2nd Lt. Brisbee, 3rd Lt. Tillinghast, Sailing Master Jones. After he enlisted he went to his fathers at Saybrook, staid four days or five, and then went to Boston and went on board the said ship & served on board that vessel for eight months during which time he sailed to Charleston, South Carolina and then on a cruise with intention to go to France for soldiers clothing. The vessel was dismasted on the Bank of Bahama when we were convoying some American vessels to the Port of Havana. He returned with said vessel to New London Conn. And was discharged at that place under same officers and no written discharge.

After he was discharged at New London he went to St. Eustatia, one of the West India Islands and there in January 1781 again enlisted on board the Brig General Washington belonging to the State of Massachusetts, carrying twelve six pounders, Jonathan Caldwell commander, don't recollect the Lieutenant. After serving on board only three days he (& the vessel) was captured and taken prisoner and was taken to Antiqua in the Port of St. Johns and imprisoned by the British for eleven months, when he was released in the month of December or November 1781 and exchanged at Philadelphia. He was in close jail for the whole of the eleven months and had but one shirt and went naked most of the time. There were about three hundred Americans in prison with him at Antiqua.

Author's note – Danial Hilyard's pension application was approved and he received $89.72 per year for his service under the 1832 Pension Act.

He passed away on the 20th of November 1834.

Isaac Squire

State of Connecticut, County of Fairfield

In the Probate District of New Milford

On this 3rd day of August A.D. 1832, personally appeared in open court before the Court of Probate for said District now sitting at Sherman in said District & County, Isaac Squire, a resident of said town aged eighty years, who being first duly sworn according to law doth on his oath make the following declaration in order to obtain the benefit of the provision of the act pf Congress passed June 7th 1832.

Author's note – I am going to skip over Mr. Squire's initial land service and jump in when he goes to sea.

That in January or February 1776 and after his return from home, he enlisted as a carpenter on board the Brig Defense of sixteen guns, commanded by Seth Harden Esq, Captain Jedking while he was on board said Brig. They captured two

English ships & a Brig in Nantasket, vessel's having on board Col. Campbells Regiment, while they took prisoner & took them into Boston & they were afterwards exchanged. They afterwards went cruising to the West Indies & took a Guinea ship which had sold her slaves & was on her return. They afterwards captured a British ship loaded with provisions and took it into Boston. He was in the service a board this Brig as near as he can recollect about one year when he returned to New London, the place from where he started.

He then a few days after enlisted on board the Brig Genl. Washington of 16 guns as carpenter, commanded by William Rodgers Esq, and was on her as near as he can recollect about one year. During which time he went cruising off the West Indian Islands and captured an Irish Brig loaded with butter & potatoes & sent it into Boston. He returned home in the latter part of the winter of 1778.

That he remained at home till the latter part of March of the same year, or in the month of April and then again enlisted on board the ship Mars of 22 guns, commanded by Gilbert Ash Esq, and went cruising off the West Indian Islands and captured one or two vessels loaded with provisions & sent them into Boston & returned home late in the fall or early December of the same year, having been absent about eith months.

Author's note – Isaac Squire's pension application was approved and he received $36.50 per year for his service under the 1832 Pension Act.

He passed away on the 29th of December 1834.

David Hawkins

Rhode Island District, County of Providence

On this 7th day of October A.D. 1833 personally appeared in open court being a Special District Court of the United States within aforesaid District, now sitting, David Hawkins, resident in North Providence in said County & District aged 76 years, who being first duly sworn according to law doth make the following declaration in order to obtain the benefit of an act of Congress passed June 7th 1832.

Author's note – I am going to skip over Mr. Hawkins initial service and jump in when he enters aboard the *General Washington.*

And he further says that in May 1778 he enlisted as a marine on board the George Washington, a Ship of War of 20 guns then lying in Boston in the service of the United States & under the command of Thomas Hardy Esq. We immediately sailed on a cruise at sea and did not get back to Boston until about 26 of August. This part he distinctly remembers having served a little more than three months and one half – He also remembers that on about the 16 of August we had engagement and severe battle with the British Ship called Lerant, a vessel of 28 Guns. But although she was superior, we sank her and out of 130 men, which she had on board, only 19 escaped & them we took on board our vessel.

Author's note – David Hawkins pension application was approved and he received $43.33 per year for his service under the 1832 Pension Act.

He passed away on the 4th of May 1840.

USS Hancock

Schooner

There were two vessel's named *Hancock* that saw service during the Revolutionary War. One a schooner and the other a frigate.

Hancock the schooner was a converted merchant vessel loaned to the Continental Congress in 1775. She was armed with 6 guns and manned with about 70 men. She sailed in the Continental Navy until 1777 when she was returned to her owner. I did find a pension declaration from a George Whipple who stated he served on the Schooner Hancock under Commodore Manley, but did not mention anything further.

Her captains were said to be Nicholas Broughton, John Manley & Samuel Tucker.

Frigate

Hancock the Frigate was built for the Continental navy and launched in 1776. She was armed with 34 guns and manned with about 300 men. She sailed under an American flag until she was captured by the British in 1781.

Her captains were said to be John Manley.

Hancock & Boston capturing HMS Fox – history.navy.mil

Moses Ayer

I Moses Ayer aged seventy one years, born in Haverhill Massachusetts, a citizen of the United States, now resident in Salem Maine, upon oath testify and declare that at Boston in March 1777 I shipped as a marine for twelve months on board the continental ship Hancock of thirty two guns, commanded by Commadore John Manley. In our cruise we captured the British Frigate Fox and several merchant vessels. I sailed on bard the Hancock until July 1777 when we were captured by the British Ship Rainbow and British Brig Cabot. I was carried a prisoner to Halifax where I was confined in prison, part of the time in irons, until November 1777 when I made my escape with two others and returned to Boston. I served in this cruise including the time I was in prison and until I got back to Boston twelve months. And according to the best of my recollection I was paid for twelve months. I took no formal discharge.

Author's note – Moses Ayer's pension application was approved and he received $8.00 per month for one year of service as a marine under the 1818 Pension Act.

He passed away the 27th of February 1823.

David White

I David White of Dorset County of Bennington and State of Vermont of lawful age testify & say that I entered on board the Hancock frigate as a marine, Capt. John Manley in the fore part of March 1777 – The Hancock sailed from Boston Harbor in company with the Boston frigate The Tartar and the Mifflin, the fore part of May 1777 – On the 7th day of July following the Hancock on board of which I then was, was captured by the Rainbow and her tender vessel. We were carried into Halifax and there remained prisoners until the 7th day of January 1778. We were then sent away from Halifax, about 280 in number under a convoy, and after being out a number of days the prisoners arose and took the vessel and run her into Marblehead – That I was in the service against the common enemy ten and eleven months.

Author's note – David White's pension application was approved and he received $8.00 per month for 9 months of service as a marine under the 1818 Pension Act.

He passed away the 1st of January 1834.

Samuel Curtis

I Samuel Curtis of Amherst in the County of Hillsborough and State of New Hampshire, Esq. aged 70 years, a native and resident citizen of the United States, do solemnly declare and say that as near as I can recollect, about the 20th day of March AD 1777 I entered the United States service as surgeon on board the Continental Frigate Hancock of 36 guns, commanded by John Manley Esq. That we sailed from the Port of Boston about the last week in May 1777. That in about five weeks after sailing we fell in with, & after an action of about 3 glasses with the British Frigate Fox, commanded by Capt. John Fotheringham, of about 24 guns.

We took her, manned her & proceeded on our cruise. That in July following we fell in with a British 74, commanded by Sir George Collier, the Flora Frigate, and a Brig Cabot. When we were compelled to submit, and were carried into Halifax, Nova Scotia, where we were confined in close prison until the beginning of January 1778. Then 300 of us prisoners were put on board the ship Royall Bounty to be sent to Newport, Rhode Island, under convoy of said Brig Cabot, to be exchanged as we were told. After being at sea 3 or 4 days we were separated by a gale of wind from the convoy and we thought a conventional opportunity had presented. We took command from the Capt. And after running many risques we got into Marblehead harbour, the latter end of January 1778.

When I returned home to Marlborough Mass., ready to be called on board some other ship; That after a lapse of between 2 or 3 years, the Hon. Nathan Gorham, then member of Congress from Mass., took my commission & procured me a final settlement certificate for $963 dollars, which I was compelled to sell for about 1/3 of its nominal value. I presume he left it in the War Office, for I have not seen it since. That I never received any regular discharge during the remainder of the war. That from bodily indisposition & severe lameness of late years, and from my reduced circumstances, I do really need the assistance of my country for support.

Author's note – Samuel Curtis' pension application was approved and he received $20.00 per month under the 1818 Pension Act.

There is no mention of his death.

Elisha Fuller

State of Massachusetts, Suffolk, Boston April 1818

I Elisha Fuller a citizen of the United States now resident at Charleston in the County of Middlesex in the State aforesaid, do on oath declare, that I served in the War of the Revolution, the full term of time required by a Law of the United States, made and passed in the month of March, A.D.1818, entitled "An Act to provide for certain persons engaged in the land and naval service of the United States, in the Revolutionary War," as one requisite t entitle me to be placed on the pension list of the United States, having entered the naval service of the United States, on the Continental Establishment, and served as a Mariner & Marine against the common enemy; from sometime in the winter of the year 1777 until sometime in the winter of 1778, when I returned home, having been absent in the service of my country one full year.

That I belonged to the American Frigate Hancock, commanded by Commodore John Manly. After being on board said Frigate about three months she was captured by the British Frigate Rainbow & Cabbot Brig of war, when I was carried to Halifax a prisoner of war. After having been detained on board said British Frigate Rainbow a number of months as a prisoner, I made my escape and returned home to Lynn, in the County of Essex, where I then lived aforesaid – When immediately shipped again & entered on board the American Frigate Raleigh, Capt. John Berry commander, which was captured three days after sailing from Boston. I also served the Frigate Trumbull & the Frigate Confederacy, on each of them eight months.

Author's note – Elisha Fuller' pension application was approved and he received $8.00 per month under the 1818 Pension Act.

He passed away the 3rd of November 1822.

Isaac Billington

I Isaac Billington aged fifty nine years a citizen of the United States, born in Middleboro Massachusetts, now a resident in Wayne, Kennebec County Maine, upon oath declare that I enlisted in December 1775 as a private soldier in the war of the revolution against the common enemy upon the continental establishment, for during the year service of 1776 – into the company commanded by Cap.t Pearce and regiment commanded by Col. Ward of the Massachusetts Troops. I continued to serve in said company & regiment until September following when permission was given for those in the land service to enter the continental naval service. I accordingly enlisted at Boston on board the Ship Hancock, commodore Manley, for the cruise as a marine under Captain Lyman, who had command of the marines on board said ship. The Hancock at the time I enlisted was fitting for a cruise at Newbury Port, & I went from Boston to that place under the command of Capt. Ayer. I continued to serve on board the Hancock and in October 1776, while on duty as a sentinel on board said ship, in attempting to prevent a marine from passing, he resisted me, a scuffle ensued and I was thrown across the ships deck striking the gang way on the carriage of one of the guns, which broke one of my ribs on my left side & dislocated my right shoulder and broke the breast bone on my right side – there was no surgeon on board the ship at the time of this accident, one was procured by Sergt. Alger, and I was carried on shore at Newbury Port & remained there two or three weeks, when I was again carried on board the ship and arrived at Boston the last day of November 1776, when I was honorably discharged by order of Commodore Manley on account of said wound

Author's note – Isaac Billington's pension application was approved and he received $8.00 per month under the 1823 Pension Act.

He passed away the 16th of December 1829.

USS Hornet

The *Hornet* was a converted merchant vessel acquired by the Continental Congress in 1775. She was a small sloop armed with 10 guns and manned with about 40 men. Not a lot is written about *Hornet's* Continental naval service. And she is listed as being captured by the British in 1777, who soon afterwards condemned her.[2]

Her captains were said to be Nathaniel Bently, William Stone, Stevens & Stutsen

I'm not sure if the historical record is incorrect, or the men who sailed her. But the pension claims you are about to read state they sailed on *Hornet* in 1778 & 1779. This would have been a year or two after she is said to have been captured by the British. Granted, these men are trying to remember back over 50 years, so there may just be a lapse in memory.

Elijah Stevens

Declaration in order to obtain the Benefit of the Act of Congress passed June 7th 1832.

State of Maine, County of Oxford

On this twentieth day of September in the year of our Lord one thousand eight hundred and thirty two, personally appeared in open court before Stephen Emery Judge of the Court of Probate & within & for said County of Oxford, now sitting at Livermore in said county Elijah Stevens a resident of said Livermore aged sixty six years on the seventh day of October next, who the following declaration in order to obtain the benefits of the Act of Congress passed June 7, 1832.

That he entered the naval service of the United States as a volunteer about the last of September AD 1778 and served on board the schooner Hornet carrying eight, four pounders on her main deck & several swivels, under the command of Captain Nathaniel Bently, with the following crew – Peter Bently, Freeman Lincoln, Cornelius Jones, Gideon Stevens and others whose names he does not recollect for the term of about five months – Said schooner sailed from New Bedford (Mass) about the last of October AD 1778, nearly a month after he entered the service as aforesaid, and cruised several months between the American coast and the West Indies, took one small prize, the name of which he does not recollect, and returned to the United States and entered the port of New London (Conn) where he was verbally discharged about five months from the time of entering the service as aforesaid – The name of one of the lieutenants who served during said cruise was Howe, the name of the other officers he does not recollect.

And the said Elijah Stevens further states that he entered the naval service of the United States as a volunteer on board the Brig Hazard and served eight months under the following officers, Captain David Williams, Lieutenant Little, with the following crew – Clark Hunter, Curtis Henry, and others , officers & marines whose names he does not recollect and sailed from Boston about the first of October 1779 some days after he entered the service aforesaid, and cruised across the Atlantic Ocean twice, took several prizes, one of which was the English ship of war Active, and returned to Boston aforesaid in May or June 1780, when and where he was verbally discharged.

The said Brig Hazard carried fourteen double fortified six pounders on her main deck and two four pounders on her quarter deck & several swivels & in the said action with the said ship of war Active, the said Hazard lost eight men killed & eight men wounded – and the said Active during the same engagement lost sixteen men killed & had sixteen men wounded, among the wounded her captain and among the killed on of her lieutenants.

And the said Elijah Stevens further states that about the last of August 1780 he shipped on board a merchant vessel at Warren (R.I.) and sailed for the West Indies where the said vessel was lost in a hurricane on the south side of Hispaniola (the crew being saved)and the said Stevens with several of said crew went on board the sloop from Virginia & sailed for some port in that state, & on our passage were taken prisoner & carried to the island of New Providence & there confined three months and then escaped & arrived in safety at Cape Francis (Hispaniola) where he enlisted as a volunteer with one Martin Estatbrook on board an American armed vessel being a Letter of Marque (name not recollected) and sailed for Boston, but was landed and verbally discharged at Chatham (Cape Cod Mass) about three months after his said enlistment, and he is unable to recollect the names of the officers or any of the crew except the said Martin Estabrook.

Author's note – Elijah Stevens' pension application was not approved because he failed to provide proof of his service.

There is no mention as to his date of death.

John Fields

Author's note – John Fields' pension application is extremely long, detailed, and he does not mention his naval service until the very end. So, I am going to start with his swearing in, which I feel also has beneficial personal information, and then skip past four pages of his land service and dive right into his tour aboard *Hornet.*

Declaration

In order to obtain the benefit of the Act of Congress passed June 7th 1832.

The State of Ohio, Greene County

On this first day of December in the year of our Lord eighteen hundred and thirty four in open court being the Court of Common Pleas for the County aforesaid now siting personally appeared John Fields, a resident of the county and state aforesaid, aged 80 years on the 14th day of May last, who being first duly sworn according to law, doth on his oath make the following declaration in order to obtain the benefit of the Act of Congress passed June 7th 1832.

This declarant states that he was born in Springfield Township County of Burlington, in the State of New Jersey on the 14th of May 1754 and that whilst residing there he entered the service of the United States under the following named officers, and served as herein stated.

This declarant was also engaged in the Naval Service of the United States during the Revolutionary War. He thinks it was sometime in the latter part of the year 1779 – but the particular date has faded from his memory entirely – That he volunteered and served the full term of three months on the ocean. The vessel in which he served was a Brig called the Hornet, commanded by Captain Stutsen , John Dillion was the First Mate, and the Second Mate was Thomas Smith. The vessel was a Privateer. This declarant embarked at Philadelphia – the Brig stood out to sea and continued cruising on the ocean for about one month when she came in contact with a British man of war and was chased by her about three days and was finally drove into a small inlet called "Tom's River" he thinks somewhere off the coast of New Jersey, but he does not know in what section or district of the state. We were, before entering this inlet fired on by the British vessel, and lost our Captain and seven men. Our vessel remained in this inlet about one week closely guarded by the British vessel – She being to large a vessel to follow us into this inlet.

Author's note – This sounds like a documented encounter *Hornet* had with *HMS Fly*.

When she sailed away. A short time after this British vessel had departed, we again stood out to sea and after sailing about for some time, we came across a British Merchant vessel on her voyage from the West Indies; We attacked her – after firing one gun she struck, and we conducted her into "Tom's River" aforesaid. Shortly after or arrival at "Tom's River" this declarant shared in the prize, left the Hornet, and returned across the country to his home aforesaid in New Jersey after serving for the full term of three months.

Author's note – John Fields' pension application was approved and he received $76.66 per year for 23 months of service in the militia, his sea service was not credited, under the 1832 Pension Act.

There is no mention as to his date of death

Joseph Whitecar

State of New Jersey, Cumberland County

On this first day of October in the year of our Lord one thousand eight hundred and thirty two, personally appeared in open court before Reuben Hunt, George Sanden, Edmund Sheppard and Jeremiah Steell, Judges of the court of Common Pleas of said county of Cumberland and State of New Jersey now sitting – Joseph Whitecar, a resident of the Township of Downe in said county and state, aged seventy seven years who being first duly affirmed according to Law, doth on his affirmation make the following Declaration in order to obtain the benefit of the act of Congress passed June 7th 1832.

Author's note – I am going to skip over Mr. Whitecar's land service and join his declaration when he gets into his sea service.

May 1, 1779 shipped on board the Sloop Hornet as a marine under command of Capt. Stevens, 1st Lieut. Mansfield, 2nd Lieut. Smith. We carried ten guns and fifty men. Sailed from Little Eggharbour, we soon fell in with and joined three American vessels called True Blue, Polly Sly, the other not recollected. While out in the Hornet took four British vessels, got three in, one carried ten guns and eighty men, another four guns, another Mugeon built Sloop loaded with molasses.

In the same year shipped on board Schooner Hawk of ten guns and eighty men. Capt. Stillwell, 1st Lieut. William Trean, 2nd Lieut. Joshua Smith. After being out two days was taken by the British and put in irons eight days and nights when we arrived in New York. Was then put on board of a prison ship. There kept till the 16th of Nov. 1779. Then exchanged. Was out this term six months and a half.

Author's note – Joseph Whitecar's pension application was approved and he received $43.33 per year for his service under the 1832 Pension Act.

There is no mention as to his date of death

USS Lee

Schooner

There were two vessels in the Continental Navy named *Lee*. The first was a merchant schooner which saw service in the Atlantic Ocean. The second was a galley which saw service on Lake Champlain, which I could find no pension claim for.

The Schooner was chartered, or loaned, to the Continental Congress in 1775, was armed with 6 guns and 10 swivels. She sailed under the American flag until she was returned to her merchant owner in 1777.

Her captains were said to be John Manley, Daniel Waters, & John Skinner.

Joshua Atkins

District of Massachusetts

On this twenty seventh day of March in the year of our Lord Eighteen hundred and eighteen, before me, this subscriber, one of the Judged of the Circuit Court of Common Pleas within the state aforesaid, the same being a Court of Record appeared Joshua Atkins aged sixty one years, resident in Boston in the Commonwealth and District aforesaid, who being by me first duly sworn, according to law, doth on his oath make the following declaration in order to obtain the provision made by the late Act of Congress entitled "an Act to provide for certain persons engaged in the land and naval services of the United States in the Revolutionary War."

That he the said Joshua Atkins enlisted for an indefinite period on the first day of August in the year of our Lord Seventeen hundred and seventy five in Boston in the Commonwealth aforesaid – Massachusetts – on board the public armed vessel, the schooner "Lee" of which Daniel Waters was Commander and Edward Davis was First Lieutenant. That he continued to serve on board her as a private seaman from said first day of August without interruption, until July in the year of our Lord seventeen hundred and seventy six; That he did not leave said vessel, but continued to serve on board of her as a private seaman until and through her next cruise, which ended in three months or more after said July, being in all, more than nine months, that he served on board said vessel without being absent from her, that while on board said vessel there were captured by her three vessels, towit a brig – a schooner – and a sloop; And that he has never received any wages, pay, prize money or other remuneration for any of the service by him rendered on board said vessel as aforesaid.

Author's note – Joshua Atkins' pension application was approved and he received $8 per month under the 1818 Pension Act.

There is no mention as to his date of death

James Richardson

State of Maine, Hancock County

On this tenth day of July A.D. 1833 personally appeared in open court, before the Honorable Jol Nelson Esquire Judge of the Court of Probate for the County of Hancock now sitting, James Richardson, a resident of Mount Desert in said County of Hancock aged seventy nine years, who being first duly sworn according to law doth on his oath, make the following declaration in order to obtain the benefit of the act of Congress passed June 7th 1832.

That he entered the service of the United States under the following named officers & served as herein stated- That he enlisted on board the United States Schooner Lee as a seaman at Salem in the Commonwealth of Massachusetts on the fore part of June 1776. That Danial Waters was Captain of said schooner – First Lieutenant names do not recollect, he thinks his name was Jarvis, the 2nd Lieutenant name was Knight – Sailing Master Dennis. That he was enlisted aforesaid by Lieutenant Knight, in a few days said schooner arrived at Salem from Boston & he went on board & sailed immediately on a cruise – was twice driven int a harbour by the Sloop of War Milford, afterwards proceeded his cruise to the Eastward & captured a British Brig (name not recollected) loaded with stores & bound to Jamacia – went on board said Brig with Lieutenant Knight, the Prize Master & returned to Boston -at Boston he again went on board the Schooner Lee, which had arrived at Boston before the Brig – again sailed on a cruise to the Eastward & captured a British schooner bound to the West Indies loaded with boards & fish & returned to Boston, where he was discharged in Dec. the same year having served at this time not less than six months & ten days.

Author's note – James Richardson goes on to detail land service, which I have omitted.

He was granted a pension and received $37.00 per year under the 1832 Pension Act.

There is no mention of his death.

Edward Jarvis

District of Massachusetts

On this fifteenth day of July 1820 personally appeared in open Court, being a Court of Record, A Special District Court of the United States for the said District, Edward Jarvis, aged sixty one years, resident in Boston in the County of Suffolk, in said District, Sailmaker, who, being first duly sworn according to law, doth on his oath declare, that he served in the Revolutionary War as follows:

To wit on the first day of August A.D. 1775 I entered on board the Continental Schooner Lee, Daniel Waters Captain, and I served as a common seaman in said Schooner without any intermission from said first day of August until July 1776. While I was in said vessel as aforesaid, two prizes were taken by us, one retaken by us, and we navigated the powder ship previously taken by Capt. Muckford into Boston Harbour. I have not received any wages or prize money for the part I took in performing these services.

Author's note – On another form in his pension application Mr. Jarvis states that his father, Edward Jarvis, was the *Lee's* First Lieutenant.

His pension application was approved and he received $8.00 per month for his service under the 1818 Pension Act.

There is no mention of his death.

USS Mosquito

The *Mosquito* was either a schooner or sloop, records differ on her type and her pension claimants refer to her as a brig. There is even a question as to whether or not there were two vessels named *Mosquito*.

She is said to have been a private vessel purchased in 1775 by the Continental Congress. She sailed under the American flag until she was burnt by the British in 1777. She was armed with 6 guns, several swivel guns, and crewed with around 75 men.

Not much information is available about *Mosquito* and I think the pension claims you are about to read here probably detail her service better than any source I could find.

Her captains were said to be Thomas Albertson & John Harris

Moses Stanly

Author's note – This declaration from Moses Stanly you are about to read was found in George Catlett's pension application. Mr. Catlett's was deceased, and his wife and children were filling a claim to obtain his benefits.

Moses Stanly did have his own pension claim folder, but this declaration was actually written much better than his statement in his pension claim, which I will mention at the end.

I certify that I was acquainted with George Catlett of Port Royal in Caroline County in the Revolutionary War – Early in the revolution, but at what particular date, I could not now undertake to say. I enlisted in the marine service of Virginia for the term of twelve months, and when the term was nearly expired, I re-enlisted for the term of three years or duration of war, and soon after my second enlistment , I think about one month, I was taken into service on board the Mosquito Brig.

The officers on board of said brig when I entered her were Captain John Harris, Lt. Chamberlayne Midshipman Alexander Moore, Alexander Dick Captain of Marines, and George Catlett Lieutenant of Marines; The minor officers were Timothy Laws, a man named Eagles, whose given name I do not recollect, William Coleman, John Brock, Robert Hamilton, and some of the privates were William Mitchel now living in this county, Thomas Cotrlli, John Dodd drummer, William Dowlins or Dollins, George Spillman, Tom Chandler, William John & Joseph Dishman and Sakin Farich – The Mosquito was at Norfolk when I entered her and a short time after I was aboard, say about one week, she set sail on a cruise, and stood away for the West Indies, with a crew of about seventy five men including all the above named, with Charles Dekay, who was Sailing Master – We captured the Snow John and afterwards the Noble and carried the last named into Point Pitre – The Snow John got away from us.

Author's note – I think he is referring to Pointe-a-Pitre in the island of Guadeloupe.

At Point Pitre our crew took the small pox, which detained us several weeks – When we were recovered of that we set sail on another cruise, and fell in with the British ship Ariadne, a vessel of superior size to the Mosquito. We surrendered to her, after having made every effort to escape, and not being able to fight her, and she took us into Barbados -The Ariadne was now commanded by Captain Collier – From Barbados all our officers were immediately sent to England, and we the seaman were put in prison at Bridgetown, Barbados – Here we were kept seven months and fifteen days, and were then sent over to England in the Antelope and were confined, some in prison ships, and others in Fortune Jail – Billy Mitchel aforesaid named was put in jail – I was put in a ship – Here we soon learned that all our officers were confined in Fortune Jail soon after arrival in England. Ralph Horn another of our seaman received a note from Capt. Dick saying that all the officers were in Fortune Jail and that he should endeavor to get all the seaman there to – Some time after getting to England, I heard that Captain Dick had made his escape, and that Mitchel and others had escaped also. – I and several others continued in confinement until the war ended and the treaty of peace was published in England – Two of our men Charles Pauldin and another died in jail. Given under my hand this 22nd July 1837.

Author's note – Moses Stanley's, or Stanly's, pension application was approved and he received $80.00 per year for 2 years of service under the 1832 Pension Act.

There is no mention as to his date of death, but it does state he died in Caroline County Virginia.

William Mitchel

Author's note – William Mitchel's declaration was also found in George Catlett's pension application. I could not find a separate pension application for William Mitchel.

The deposition of William Mitchel of Caroline County, taken before me the subscriber, a justice of the peace for said County – to be read in evidence in the claim of the late George Catlett heirs, for compensation for his service in the Revolutionary War – The deponent, being first duly sworn, states, that he enlisted in the latter part of the year 1775 in the naval service of Virginia, for the term of one year, and served out that term in the sloop Defiance, commanded a part of the time by Capt. Callender- On the expiration of that term he enlisted again in the marine service for three years during the war, and joined the company of Capt. Alexander Dick, in which George Catlett was Lieutenant – He states that a short time after his seemed enlistment, and he thinks in the spring of 1777, that Capt. Dick's company was put on board the Brig Mosquito, and ordered on a cruise – The officers of the Mosquito, were Capt. Harris, Capt. Dick, Lt. George Catlett, LT. Chambrlayne, Midshipman Moore, Steward John Brock, Gunner's Mate Robert Hamilton, and Joseph Warrick Pilot or Pilots Mate –

the last 5 from the Defiance where I had served with them – Some of the privates were John, Joseph, & William Dishman, Moses Stanley, George Spillman, Thomas Cattrill, John Dodd, William Dollins, & Tom Chandler.

In the spring of 1777 we sailed from Norfolk on a cruise and bore away for the West Indies – We captured the Snow John, which got away from us – We captured also the ship Noble, both loaded with clothing & provision for the British army – The latter we carried into Point Petre in the Island of the Antilles and sold her.

Author's note – As mentioned in the previous claim by Moses Stanly, this would be Guadeloupe.

From the crew of the Noble we took the small pox, which detained us some time at Point Petre – After recovering from the small pox we set sail on another cruise, and soon spied another sail which turned to be the British ship Ariadne. She was a larger ship then the Mosquito & mounted more guns – We could not escape her, and we were compelled to surrender. She took us into Barbados, where the privates were confined as prisoners for seven months and fifteen days. When we were sent over to England and confined, some in Fortune Jail and others in prison ships – After a confinement there of thirty two months or thereabouts, myself and fifteen of the privates made our escape by undermining the jail – We seized a small sloop which was near and crossed over to France, whence we soon got a passage to the United States. Capt. Dick made his escape some time before me & returned to the United States. When I escaped I left Lt. Cattell in jail, in another apartment, and he did not escape till some time after me, as I understood – He returned to the United States not long before the surrender at York and joined the infantry as Capt. Dick did also- There being no naval command for them – On my return I reported myself to Capt. Dick in Fredericksburg and was ordered by him to join the infantry, which I did, and served in the company of Capt. Wallace until the surrender at York

William Dishman

State of Kentucky, County of Barren

On this 20th day of August 1832 personally appeared before the County Court of the County of Barren, William Dishman a resident of the county of Barren in the said State of Kentucky aged 77 years, who being first duly sworn according to law doth on his oath make the following declaration in order to obtain the benefit of the provision made by the Act of Congress passed June 7th 1832; That he enlisted in the army of the revolution in the year 1776 with Capt. Alexander Dick and served in (he believes) the third regiment of the Virginia State Line under the following named officers & for the time & in the manner following; he first entered the marine service in the year 1776 for two years under the said Captain Dick, and was stationed at Tappahannock (or Hobb's Hole) in Essex County Virginia in which county he then resided – after having kept guard at Tappahannock for sometime, waiting for a vessel, it was at length said that none of the public vessels belonging to the state of Virginia were fit for sea, and some of the officers had letters purporting that those marines stationed at Tappahannock should be put on board the row gallies at York (or some where in the Chesapeake) unless they would enlist in the regular army for three years – The said William Dishman, with others then enlisted for three years in the regular army under the said Capt. Alexander Dick of Virginia Line and (he believes) the third regiment. George Cattell was First Lieutenant & Charles Thornton Second Lieutenant of the company – the names of the other officers are not recollected by this said Dishman, nor does he recollect the name of the Colonel or other field officers of the regiment. He, the said Dishman was afterwards ordered to rendezvous at Fredericksburg and went under the command of Capt. Dick on board the Brig Mosquito to sail on a cruise against the enemy – He does not recollect that any of the other officers belonging to the company went on this expedition. John Harris was the Captain of said Brig & Bird Chambers First Lieutenant & George Chambers 2nd Lieutenant – After embarking on board the Brig they dropped down to Hampton Roads where they lay until the 27th February 1777 when they sailed on her passage to the West Indies – The Brig captured a transport ship (name not recollected) belonging to the enemy loaded with provisions, candles & carried her to Point Piter, Guadalupe. The said Dishman afterwards understood the captured vessel and her cargo were sold, but he never received any of the prize money – The said Dishman afterwards sailed on board the said Brig but after a few days discovered that one of the crew had the small pox, she was obliged to return to Guadalupe inoculate the rest – After they got well she again sailed, but after a few days she was captured by a 20 gun vessel the Ariadne) belonging to the British navy & all on board (said Dishman among them) made prisoners – This was to the best of said Dishman's recollection, about the 4th of June 1777 in the night – The said Dishman further states that the prisoners & he among them were carried to Barbados where they were imprisoned (he thinks at a place called Bridgetown) & detained until January 1778 when a fleet of British merchantmen under convoy touched at Barbados & took the prisoners on board, separating them & distributing them in different ships, Capt. Dick & the commissioned officers he understood they were sent to England. The said Dishman & two of his brothers, James & John (both now dead) were put on board the same vessel and came to the Island of Jamacia where they were set at liberty. He and his brothers remained on the island sometime & at length worked their passage to America on board a British merchant ship commanded by Captain Smith, bound to Philadelphia; Then in possession of the British; And by the intervention of a merchant on board, a Scotsman by the name of Buchannan, they was landed on the Delaware a short distance below Wilmington, about the month of May 1778. After landing he and his brothers went to the quarters of General Smallwood at Wilmington – he does not recollect to have seen him, but Major Smith gave them a passport to return by way of the Head of Elk and Baltimore to Virginia.

Author's note – William Dishman's pension application was approved and he received $80.00 per year for 2 years of service under the 1832 Pension Act.

He passed away on the 4th of December 1833.

USS Oliver Cromwell

The *Oliver Cromwell* was a corvette, ordered to be built by the Continental Congress and launched in 1776. She was armed with 20 guns and held a crew of around 180 men.

She was captured by the British in 1779.

Her captains were said to be William Coit, Seth Harding & Samuel or Timothy Parker.

Oliver Cromwell – Connecticuthistory.org

Ruben Godfrey

State of Connecticut, County of New London

On this 22nd day of August 1836 personally appeared before the Court Probate for the District of Norwich in said state & county in open court now in session, Ruben Godfrey, a resident of the town of Norwich in said Probate District, in said county and state, aged seventy eight years, who being first duly sworn according to law doth on his oath make the following declaration in order to obtain the benefit of the provision made by the act of Congress passed June 7th 1832.

That he entered the service of the United States under the following named officers, and served as herein stated. That on or about the 8th day of July 1776 at Nantucket in the State of Massachusetts, he was enlisted by Capt. William Coit as a seaman, to serve on board the sloop of war Oliver Cromwell of twenty guns, a state vessel, belonging to the State of Connecticut and in the service of the United States, and then lying in the harbour of New London in the state of Connecticut and commanded by said Captain William Coit.

Author's note – In a clarification letter to his original declaration, Mr. Godfrey states that Capt. Coit came to Nantucket and personally recruited Mr. Godfrey and several of his friends.

He entered and served on board said sloop of war, the Oliver Cromwell for the term of ten months and eleven days under said Capt. William Coit, when said Capt. Coit was discharged from the command of the said ship. After Capt. Seth Harden took the command of her, one month as Stewards Mate. Malally was 1st Lieutenant of the Oliver Cromwell, John Chapman 2nd and John Prentice 3rd Lt. That ship did not go to sea while the declarant served on board of her, she was fitting for sea when he entered on board of her and when ready for sea, she attempted to get to sea, but she capsized before she got to Montauk Point, she returned into the harbour of New London. Her masts were taken out and the ship was refitted for sea. But before she sailed under the command of Capt. Harden, the declarant left her and he declares that he served in the service of the United States on board said ship Oliver Cromwell previously stated ten months and eleven days.

And the declarant further states that at said New London in the year 1777, the month he cannot state with accuracy, Nathanial Shaw Esq, enlisted him to serve on board of the United States Frigate Rolla, then lying at Boston.

Author's note – As previously stated, Rollo is referring to *USS Raleigh.*

He went to Boston for the purpose of going to sea in the Rolla, but when he arrived at Boston the Rolla not being ready for sea, he went onboard the United States Frigate Warren of thirty two guns, commanded by Capt. John Hopkins. Enos Hopkins was First Lt. of the ship, Joshua Hempstead was Sailing Master of the Warren. He went to sea in the Warren in the service of the United States, he served on board the Warren as a seaman, as a topman, for the term of seven months. He made two cruises in her on the Grand Banks and South Sholes & he took several prizes while he served on board her.

Christopher Vail was on board the Warren, I recollect him when we conversed last Saturday, that he did not recollect me, I being stationed in the main top and not having much acquaintance on deck & he being in the ship but one cruise.

Author's note – Rueben Godfrey's pension application was approved and he received $24.00 per year for his service under the 1832 Pension Act.

There is no mention of his death.

Asa Pease

Commonwealth of Massachusetts, County of Plymouth

On this thirteenth day of November 1832 personally appeared in open court before Hon. Wilkes Wood, Judge of the Court of Probate for said County now sitting at Rochester within and for said County of Plymouth, Asa Pease, a resident of said Rochester aged seventy eight years having been born on the Island of Marthas Vineyard in said Commonwealth, on the twenty eight day of July 1754, who being first duly sworn according to law doth on his oath make the following declaration in order to obtain the benefits of the act of Congress passed June 7th 1832.

Author's note – I am going to jump over his initial land service and pick up at his sea service.

In the year 1777 he thinks, he cannot say with certainty, he entered as carpenters assistant on board a vessel of war in the national service, commissioned he thinks by the State of Connecticut called the Oliver Cromwell for a term of four months, Capt. Seth Harding, he thinks of Cape Cod, First Lieu. Parker, he thinks of the State of Connecticut, Second Lieu. Smith, a man by the name of Rue, he thinks an Englishman Sail Mate, a man by the name of Fresbee Sail Master, and a man by the name of Marvin he thinks of Connecticut Chief Carpenter. The ship carried twenty guns. She sailed from Acushnet River to New Bedford, where she lay when he enlisted. The ship then proceeded to sea and cruised off the coast of England and made three prizes, one of them a vessel of war called the Waymauch, which he was put. The Oliver Cromwell and Waymauch arrived in Boston the same day, at which place he and he believes the rest of the crew were discharged.

Author's note – Asa Pease's pension application was approved and he received $42.50 per year for his service under the 1832 Pension Act.

There is no mention of his death.

Ivory Snow

Author's note – Ivory Snow's declaration was found in Asa Pease's pension folder, vouching for Mr. Pease's service.

I Ivory Snow of Swansey in the County of Cheshire & State of New Hampshire of lawful age give evidence do testify & say that on the fourth day of June in the year of our Lord one thousand seven hundred & seventy seven I shipped on board the Oliver Cromwell, man of war, in the United States Service in the Revolutionary War in capacity of marine and served on board said man of war five months, that the name of the Capt. was Seth Harding, & Lieut. Timothy Parker, the Second Lieut. Smith, First Mate Rue, the Sailing Master Fresbee, the Boatswain name Pool, the Boatswain Mate Higgins, the given names of the forgoing not recollected. The Prize taken were three, one Brig & two ships. The name of the Brig was Honour from Cork bound to Halifax, the first ship taken was from Quebeck bound to Jamacia, the name of her was Resolution mounting ten guns, the other ship was from Jamacia bound to Whitehall mounting sixteen guns. Kept in company with the prize ship till we arrived in Boston – That we had no written discharge – that Asa Pease acted on board said man of war in capacity of assistant carpenter. We shipped at new Bedford Mass and were discharged in Boston Mass.

Author's note – Ivory Snow did have his own pension application and he was approved and he received $56.66 per year for his service under the 1832 Pension Act.

The declaration in his own pension folder was unreadable.

There is no mention of his death.

Enoch Crowell

I Enoch Crowell, aged fifty eight years, born in Yarmouth in the County of Barnstable and State of Massachusetts, now of Hallowell in the District of Maine, upon oath testify and declare –

Author's note – Mr. Crowell's declaration is long, so I am going to skip over his land service and get right into his time on board the *Oliver Cromwell.*

In March 1778 I entered as Quarter Master on board the ship Oliver Cromwell of twenty guns, belonging to the State of Connecticut, commanded by Capt. Samuel Parker of Norwich Connecticut for a cruise of six months against the enemy, which period I served out – In the cruise we took from the enemy two ships, Letters of Marque – one called the Admiral Keppel of twenty guns – and one called Cyrus of eighteen guns – which we brought into Boston.

I entered as a Quarter Master on board the Continental ship Lady Washington in March A.D. 1781 of twenty guns for six months – In our first cruise we were taken by the Chatham fifty gun British ship, between three and four months after I entered the Lady Washington, which was commanded by Capt. Walker of Newport Rhode Island – After I was taken, I was kept on board the enemy's prison ship at New York about three months before I was exchanged. This was my last service.

Author's note – Enoch Crowell's pension application was approved and he received $8.00 per month for his service under the 1818 Pension Act.

There is no mention of his death.

Samuel Buffum

State of Rhode Island, Newport

On this First day of October AD 1832 Personally appeared in open Court before the Court of Probate, now holden in and for said town of Newport, Captain Samuel Buffum, a resident in said Newport, aged seventy seven years, who being first duly sworn according to law, on his oath, declares as follows in order to obtain the benefits of the Act of Congress passed June 7 1832.

That he entered in the naval service during the Revolutionary War on bard the Oliver Cromwell a twenty gun ship, twenty nine pounders, fitted and commissioned by the State of Connecticut, the exact time he does not now recollect, but it was soon after Rhode Island was garrisoned by the enemy. He enlisted at Boston as a Midshipman, and served as such during the cruises of said ship, which lasted about twelve or fifteen months; Timothy Parker of Norwich Connecticut was commander of said ship, took several prizes, some of which arrived safe in Boston; was dismasted during said cruise between the Bahama Bank and Cape Florida, and returned to New London under temporary masts and took one prize while in this situation, which prize arrived I believe safe in New London.

Had a severe engagement early in the cruise with a Letter of Marque carrying twenty two guns, named Two Brothers, lost a good many men, but carried our prize which arrived safe in Boston, John Tillinghart Third Lieutenant being Prize Master. Soon after the termination of the cruise of the Oliver Cromwell, entered at Boston on board the ship Protector owned by the State of Massachusetts; commissioned by Governor Hancock, entered as a midshipman and served in her about one year. The Captains name was John Foster Williams, the Lieutenants Little and Weeks, Christian names not recollected. Larson, Christion name not recollected, Sailing master; This ship carried twenty guns, twelve nines, took four prizes, one to the leeward of Martinique, a sloop; another off Charleston of two and thirty guns, a Letter of Marque, not fully manned, bound for Jamacia. She did not change a shot with us. The third prize was a valuable recaptured Dutch Brig, at the time said to be the first Dutch recaptured prize after the declaration of war by England against Holland. The fourth prize was a Brig called Good Design, took her considerably to the northward of Cape Hatteras, she was laden with Tenerife wine bound to New York; of this Brig I was appointed Prize Master, and arrived with her safe into Boston on the 10th of May 1781. This Prize was saved under peculiar circumstances, for very soon after I had taken charge of her, two enemy Frigates hove in sight, gave chase to, and engaged us, captured one ship, the Protector. One of the Frigates was dispatched after the Good

Design, but with a strong fair wind, with all sail set, I ventured over the shoals of Nantucket and evaded the British Frigate, and as before said arrived safe in Boston.

Author's note – Samuel Buffum's pension application was approved and he received $144.00 per year for his service under the 1832 Pension Act.

There is no mention of his death.

Samuel Spencer

State of New York, City & County of New York

On this thirteenth day of August one thousand eight hundred and thirty six personally appeared in open court before the Judges of the main court of the city of New York, now sitting, Samuel Butler Spencer, now in the City of New York, a resident of Hartford in the State of Connecticut, aged seventy six years and upwards who being first duly sworn according to law, doth on his oath make the following declaration in order to obtain the benefit of the act of Congress passed June 7th 1832.

That he entered the service of the United States under the following named officers and served as herein stated, that is to say; I was born the 4th day of March 1759 at Hartford in the State of Connecticut. During the War of the Revolution and in the latter part of the year 1777, I shipped as a hand on board the State Ship called Oliver Cromwell of 26 carriage guns and

4 howitzers, commanded by Captain William Coit, at New London in the State of Connecticut, at a bounty of thirty dollars and a share and a half of Prize Money. And I served on board said ship about three years – until she was taken by the British in the fall of the year 1780 or 1781. John Smith of Hartford aforesaid was First Lieutenant of the said ship, and Berzelia Beebee of Stonington Point, Connecticut was Second Lieutenant. He died at Newbaugh in the State of New York about twenty years ago. Said John Smith is also dead, but I do not know or recollect when he died.

Captain William Coit served but a few months, and was succeeded in the command of the Oliver Cromwell by Captain Hardy of Boston, who was the commander when the said ship was taken by a British Squadron. We sailed on a cruise on the western ocean and during the cruise we took two Jamacia ships belonging to the enemy loaded with sugar and carried them into New Bedford as prizes and as such they were sold there. We arrived at New Bedford with the prizes a few days before the day called the dark day which happened in May 1780, as I think. We sailed from New Bedford to New London, from which place we were in and out continually. We captured a Kings Schooner of 10 guns off Block Island and carried it into New London and made a privateer of it. This vessel had been fitted out at the City of New York.

When the British came upon an expedition from New York to New London we ran up the River about nine miles, where we run our ship ashore and leaving a sufficient number to guard the ship. Sixty two volunteers of the ships crew, of whom I was one, being provided with arms by Capt. Hardy we marched under his command down to New London to fight the British, who had landed there in a expedition from the City of New York under the command of <u>*Traitor Arnold*</u>. *We had much hard fighting that day with the British, who after burning the town and taking Forts Griswold and Trumble, left about sun set and returned to their ships. We next day buried about five hundred men killed in the Battle.*

I served on board the Oliver Cromwell from the time she was first fitted out until she was captured by a British Squadron and carried into New York in 1781. I served about three years on board the Oliver Cromwell and was on board when it was taken by the enemy. I was made a prisoner of war and put on board the Jersey Prison Ship at the City of New York. I remained a Prisoner on board said ship about six or seven months and was then taken out by Lieutenant McKinsey of the British Army upon my consenting to serve him as a servant. I was in his service about four or five moths when he was ordered south and I being unwilling to go with him, he spoke a good word for me to Col. De Laney of the British Army stationed at the City of New York, who gave me a Pass-Port by which I passed the British Guards at the City of New York and returned home to Hartford Connecticut.

Author's note – Samuel Spencer's pension application was approved and he received $96.00 per year for two years of service under the 1832 Pension Act.

There is no mention of his death.

USS Providence

Providence was a merchant sloop which was first known as *Katy* when she was commissioned by the State of Rhode Island in 1775. Later in that same year she was brought into the Continental Navy and renamed *Providence.* She was armed with 12 cannons, 14 swivel guns and held a crew of around 60 men. She was lost in 1779 when her crew destroyed her to prevent being captured by the British.

Her captains were said to be Abraham Whipple, John Hazard, John Paul Jones, Hoysted Hacker & John Rathbun

Providence – John Mecray

Esek Whipple

State of New York, St. Lawrence County

Esek Whipple formerly of the town of Gloucester in the County of Providence and State of Rhode Island, and now of the town of Dekalb in the County of St. Lawrence and State of New York being duly sworn doth depose and depose before the Honorable Nathan Ford first judge of the Court of Common Pleas in and for the said County of St. Lawrence as follows – That this deponent during the war of the revolution was a mariner in the naval services of the United States for a longer term of time than nine months & that by reason of his reduced circumstances in life he has need of the assistance of his country for support.

And this deponent further saith that in the month of May or June in the year of our Lord one thousand seven hundred & seventy seven this deponent shipped and enlisted at the port of Bedford in the State of Massachusetts on board the sloop Providence a public vessel of war in the service of the United States commanded by John Rathbone carrying ten guns, four six pounders & six four pounders that this deponent remained on board the said vessel doing his duty as such mariner from the time of his aforesaid enlistment until the termination of her next cruise, a period of between three & four months when he was, on the return of the said vessel into the aforesaid port of Bedford regularly discharged.

And this deponent further saith that about the month of March in the year of our Lord one thousand seven hundred and seventy eight this deponent shipped and enlisted at the port of Providence in the State of Rhode Island as a mariner or sailor on board the ship Providence a public vessel of war in the service of the United States commanded by Abraham Whipple carrying thirty guns, that this deponent remained on board the said last mentioned vessel doing his duty as such mariner from the time of his last aforesaid enlistment a period of about nine months according to the best of his recollection & belief during which time said ship went to France first entering a port of that country which this deponent thinks was called Ruboef, thence proceeding to L'Orient & from thence to Brest from which last port the said vessel returned to the United States in company with the ships of war in the United States service Boston & Ranger & entered with those vessels the Port of Boston where this deponent was regularly discharged sometime near the end of the last mentioned year.

And this deponent further saith that in the month of May in the year of our Lord one thousand seven hundred & seventy nine this deponent shipped and enlisted at the port of Boston in the State of Massachusetts as a mariner or sailor on board the ship Queen of France a public vessel of war in the service of the United States commanded by the aforesaid John Peck Rathbone carrying twenty guns, that the said last mentioned vessel sailed on a cruise a short time after this deponents last aforesaid enlistment, this deponent being on duty on board the said vessel as such mariner in company with the ship

Providence, Capt. Abraham Whipple commander & the ship Ranger Capt. Simpson commander, and that the said vessel on the said cruise captured a sail of the Jamaican fleet, eight of which were safely brought into port, that the said vessel the Queen of France at the termination of her said cruise returned into the port of Boston about the month of September in the last mentioned year. That this said vessel on or about the month of November in the year last aforesaid sailed again (this deponent being still on board such vessel as such mariner) to Charleston the capital of the State of South Carolina where shortly after the arrival of said vessel she was sunk in the channel in order to obstruct the passage of the British fleet & this deponent transferred with the rest of the crew of the aforesaid vessel to one of the forty on the river Ashley where this deponent continued on duty until the surrender of Charleston to the enemy & was put on board a prison ship at Charleston 7 in June then was sent a prisoner of war to Philadelphia where this deponent was exchanged and received a passport to return to Providence to which place he returned in July the next where he was discharged in the last mentioned month or in the proceeding month of August.

Author's note – Esek Whipple's pension application was approved and he received $8.00 per month under the 1818 Pension Act.

There is no mention of his date of death.

Vail Richmond

Declaration in order to obtain the benefit of the Act of Congress passed June 7th 1832.

State of Connecticut, County of New Haven

On this 27th day of August 1832 personally appeared in open court before the Probate Court of Guilford district, now sitting Vail Richmond, a resident of the town of Madison, County of New Haven and State of Connecticut aged 70 years who being first duly sworn according to law doth on his oath make the following declaration in order to obtain the benefit of the Act of Congress passed June 7th 1832.

That he entered the service of the United States under the following named officers and served as herein stated.

That he was born in Providence in the State of Rhode Island May 12th 1762 according to tradition received from his parents, having no record or other written documents to show the date. That about the year 1779 he removed to Killingworth in the county of Middlesex where he resided till about the year 1800 when he removed to that part of Guilford in the County of New Haven and state of Connecticut which now constitutes the town of Madison in said county & state where he has lived till the present time.

That as near as he can recollect in the year 1778 he enlisted as a private sailor on board the Continental Ship Providence of thirty guns Commadore Andrew Whipple Commander, 1st Lieutenant Pitcher, 2nd Lieutenant Duvall, 3rd Lieutenant Hopkins, 4th Lieutenant Fletcher, Sailing Master Goodwin, Captain Craig of Providence aforesaid formerly in the merchant service was an acting midshipman on board the ship and was the recruiting officer under whom he enlisted. Dispatches for Dr. Franklin were on board the ship in charge of Captain Jones, Dr. Franklin was taken to France.

Autor's note – I think he is referring to just dispatches being taken to Dr. Benjamin Franklin, America's Ambassador to France, not that Dr. Franklin was on board the *Providence.*

The British blockading squadron lay below consisting of one fifty gun ship, one called the Lark of thirty six guns (the Lark was said to be an old East India man formally a Merchantman, now mounting 36 four & six pounders) and several smaller vessel alarm boats were stationed by the enemy from the squadron almost to Providence. In February of said year he enlisted and in March they sailed, about one month from the time of his enlistment in the night. The alarm was given by the boats by burning of dry & wet powder at their mast heads, but the wind being fair and brisk they came down on the Lark which was moored across the channel before the other ships could come to her assistance and after a short but severe action they cleared her and went to sea, giving the fifty gun ship which lay below and was getting underway a broadside in passing. They were chased three days by her but escaped. And got safe to Brest on the 15th of the following May according to the best of his recollection carrying in as prize a Brig from & headed to Ireland. They lay at Brest about three months, made some repairs and by order of Dr. Franklin sailed on a cruise from the Bay of Biscay to the banks of Newfoundland. They captured three vessels, one a Brig from London for Gibraltar laden with provisions, the second a Brig from Jamacia for London with rum, the last a Snow with fish from the Grand Banks for some port unknown. In December following they came into Boston and he was discharged having been about ten months. While they lay at Brest he saw Captain John Paul Jones who was there in the Ranger with his prize the Drake. He was on board the Providence a number of times. The dispatches carried by Captain Whipple in the ship Providence were fastened to an 18 pound shot and placed on the Quarter Deck during the action and chase which followed.

Author's note – The dispatches were more than likely attached to the shot in case they were captured. They would be thrown overboard and carried to the bottom by the shot. Hence, not falling into the hands of the British.

Vail Richmond's pension application was approved and he received $46.66 per year for his service under the 1832 Pension Act.

He passed away on the 20th of April 1849.

Othniel Brown

State of Connecticut, Tolland County

On this 17th day of September 1832 personally appeared in open court Rodolphus Woodworth Esq. Judge of the Probate for the District of Stafford, holding the Court of Probate for said District now sitting the same being a court of record, Othniel Brown of Stafford in the County of Tolland in the State of Connecticut aged seventy three years who being first duly sworn according to law doth on his oath make the following declaration in order to obtain the benefits of the Act of Congress passed June 7ht 1832.

Author's note – Othniel Brown's pension application starts out with a very lengthy detail of his land assignments, which I am going to skip over. So, I am just going to jump into when he joins the *Providence.*

September 1779 when left Gloucester and went to the town of Boston in the state of Massachusetts and there at said Boston enlisted into the naval service of the United States as a marine, under Capt. Jones, then of providence in the state of Rhode Island (who is now dead) He, said Jones, was a Capt. Of marines on board the ship Providence commanded by Commadore Abraham Whipple, who then commanded a squadron which was then stationed at said Boston, which consisted of four ships including the Providence, the names of the other three ships and the names of the officers that *belonged to them at that time I do not at this time recollect. Immediately after my enlistment as a marine I went on board the ship Providence which was the said Commodore Whipple's ship, after which the said Commodore Whipple with the squadron under his command remained at said town of Boston two or three weeks for the purpose of enlisting more men, I then sailed in the ship Providence with the said Commodore Whipple on board of the same and in company with the three other ships on a cruise at sea. I remember that the squadron took several prizes after we sailed on said cruise from said Boston and they were sent into port. And the said four ships became much damaged in consequence of a severe gale which we experienced at sea and which lasted four days. The said Commodore Whipple ordered all four ships to sail to Charleston in South Carolina for the purpose of repairing said ships and accordingly said squadron consisting of said four ships sailed immediately to said Charleston. Within two or three days as I think after said squadron arrived at said Charleston, said town of Charleston as well as Commodore Whipple's squadron were completely blockaded by twelve British ships of the line. If I rightly recollect in this perilous situation the officers, marines, and sailors that belonged to Commodore Whipple's squadron left the same, except a few who remained on board as sentinels, and were stationed in a fort at said Charleston until said fort surrendered to the British forces, which I think took place on the 12th day of May 1780. While we were quartered at said fort myself and many of the marines which belonged to the said Commodore Whipple's squadron frequently went on to the lines and fought the British* *forces at said Charleston on land. And after said fort had surrendered the said Commodore Whipple and the other officers that belonged to his squadron went home on parole and the marines and sailors that belonged to said squadron were left at said Charleston to be exchanged. But instead of being exchanged as agreed, we were on the 13th day of said May put on board British prison ships. I was myself with the rest of the marines and sailors that belonged to the said Commodore Whipple's said ship Providence put on board of one of said British prison ships where we were kept confined about six weeks on board of the same, and during said six weeks we were reduced almost to a state of starvation, as we had nothing furnished us fit to eat and many times it seemed to me that I should die for want of food. I will not attempt to further to delineate my sufferings for the want of food while on board said prison ship. Suffice it is to say that my sufferings were great as well as those of my fellow prisoners on board said prison ship and when we solicited the British officers for food we were told by them that if we would enlist into their service and go on board their ships we should fare well, if not we should suffer the consequences. At one time we were taken from said prison ship and put aboard a British 74 gun ship and there kept two days and had during said two days a plenty of good wholesome food to eat and grog to drink. We were then remanded back to said prison ship and told that if we would enlist into their British service we should live as well as we had during said two days that we were on board said British 74 gun ship. The British officers used all the means in their power to induce us to enlist and go on board of their ships, but we had previously determined that we would not enlist into the British service if we died for want of food, and we remained steadfast in that resolution. And the British officers finding that they could not enlist us into their service by persuasions, threats, and starvation, at length put us in a vessel called a cartel to Chester, about sixteen miles below Philadelphia in Pennsylvania where we were landed and exchanged.*

Author's note – Othniel Brown's pension application was approved and he received $60.54 per year for his 8 months & 21 days of service under the 1832 Pension Act.

He passed away on the 28th of September 1843

Cornelius Arey

I Cornelius Arey of Providence in the District of Rhode Island on solemn oath do declare and say – That in the beginning of the Revolutionary War I enlisted as a private soldier in Col. McDaugles regiment in the Continental army of the United States & in the New York Line, the name of the Captain of our company I do not recollect – in which service I continued Eighteen months – I marched to Canada in said regiment – was with General Montgomery in his life guards at the capture of St. Johns, & afterwards at the taking of Montreal and at the storming of Quebec when my said General was killed. Sometime after I left the land service & in November 1779 entered aboard the Frigate Providence a vessel then in the service of the United States, said vessel was under Commodore Whipple, and sailed to Charleston South Carolina. A little before Charleston I was put on board the Queen of France & from thence was put into the fort. During the siege I lost one of my eyes in an engagement with the enemy & in consequence of which was confined in the hospital eleven months.

Author's note – Cornelius Arey's pension application was approved and he received $8 per month for his service under the 1818 Pension Act.

There is no mention of his date of death.

Benjamin Camp

State of New York, Otsego County

On this 16th Dy of October 1832 personally appeared in open court Benjamin Camp of the town of Hartwick in said County who being duly sworn in open court it being said that he enlisted on board the frigate Providence as a mariner in the year seventeen hundred & seventy seven in the month of May or June. Said frigate was commanded by Commodore Abraham

Whipple, this deponent enlisted at Providence, Rhode Island. The frigate then sailed for Penn Bief in France & lay there about three months to repair & then sailed for Providence with army ammunition & clothing for the American soldiers. On our return we captured four ships and arrived at Providence about the middle of the year '78. In the latter part of the year '78 was in company with a ship called the Queen of France & the Ranger, had a battle with the Jamaican Fleet and took her at the banks of Newfoundland. Took the fleet and brought it to Boston. Started again & took the ship London & before I left the service we captured five other vessels.

In the latter part of the year '80 I went on board the ship Charming Sally and was taken prisoner at the harbour of New Port by the British ship Triton & put on board the prison ship Scorpion where I stayed aboard three months, being exchanged for British prisoners at Elizabethtown, New Jersey. The whole time I was in the Continental Service was about 2 years.

Author's note – Benjamin Camp's pension application was approved and he received $8 per month for his service under the 1818 Pension Act. He then applied under the 1832 Pension Act and received $98.00 per month for 2 years of service.

He passed away the 12th of October 1860.

I chose to use his 1832 pension application because he includes his service on the *Charming Sally*, which he omitted in his 1818 pension claim.

USS Queen Of France

The *Queen of France* was a merchant ship purchased from the French. She was fitted out as a frigate and armed with 28 guns. After her purchase from France, she sailed under an American flag until being sunk in 1780.

Her captains are said to be Joseph Olney & John Rathbun

John Peck

State of Kentucky, City of Lexington

On this 12th day of December 1833 personally appeared before C. Hunt Esq. Mayer of the city before said, in open court, John Peck a resident of said city, who will be 64 years in January next, who being first duly sworn according to law doth on his oath make the following declaration in order to obtain the benefit of the Act of Congress of June 7th 1832.

That he was born in the town of Boston on the 12th January 1770 as appears by a genealogical statement now in his hands, taken from the records of his fathers family, which have been long since lost or mislaid. He remained a citizen of

Massachusetts until 1816 when he removed to Lexington K. where he has resided ever since. He was living in Boston in 1779 when he entered as Mizen Top Boy on board the continental frigate the Queen of France, Capt. John Peck Rathbourn commander. After remaining for some time on board she sailed on a cruise and in company with the Providence of 32 guns, Com. Whipple, and Ranger of 18 guns, Capt. Simpson. His father was at the same time on board the Queen of France. After a cruise of nearly two months, whilst lying on the banks of Newfoundland in a thick fog which continued for nearly three weeks, about meridian they were surprised at hearing guns and bells sounding, as from a large fleet; and the fog just then clearing up the American ships were found to be in the midst of a fleet of from 150 to 200 sail as reported by the prisoners, under convoy of a 74 and several frigates of the enemy. Captain Rathbourn passing himself as a Captain of a British frigate, captured one of the enemy's merchant ships without giving the alarm, by which he obtained the private signals of the enemy. Com. Whipple gave the signal to stand out of the fleet, but on the urgent application of Captain Rathbourn, he permitted him to remain and make as many captures as possible. He succeeded in capturing as many as five vessels before next morning without giving alarm, whilst the other two vessels had captured 5 0r 6 more, in all 10 or 11 vessels, one of which was a Snow, the rest were ships. The next morning suspicions being exerted, and the Queen of France not having any more spare hands, she stood out and with the other vessels were chased by a 44 frigate until knight, when the enemy gave over the chase. Of the prizes, eight reached port safely, the rest were recaptured. Those prizes were very valuable, being loaded with West India produce, with many passengers, and all having more or less arms & ammunition, some as many as 26 guns. His service on board the Queen of France continued at least six months. He annexes here to an evidence of his service as above stated, a copy of a private account & receipt showing the share of the prize money which his father received on his account, as paid at Boston. On the certificate of the Notary Public annexed thereto, there is a mistake which represents the Queen of France as a private armed vessel, when she was in fact a continental vessel.

After return to port he again entered into various privateers and letters of Marque –

Author's note – Letters of Marque were commissions authorizing privately owned ships, known as privateers, to capture enemy merchant vessels.

In which he served to the end of the war, having during the war been twice captured by the enemy. The first time on board the Nancy of 18 guns, Capt. John Hopkins, and was carried into New York and put into the Jersey prison ship whence he escaped by swimming after three weeks imprisonment. The second time on board the Waxford, and carried into Krigsale and after remaining there after a few weeks he escaped to France with some French prisoners who were returning home being exchanged.

Author's note – John Peck's pension application was approved and he received $20.00 per year for his service under the 1832 Pension Act.

He passed away the 31st of May 1847.

Henry Skinner

Eastern District of Pennsylvania

On this 19th day of March AD 1819 before the transcribing Judge of the District Court of the United States in the Eastern District of Pennsylvania, personally appeared henry Skinner aged sixty eight years resident in the said District who being duly sworn according to law doth make the following declaration in order to obtain the provisions made by the late Act of Congress entitled "An Act to provide for certain persons engaged in the land and naval services of the United States in the Revolutionary War. That he the said Henry after having been in the service of the State Of Massachusetts on board the Tyrannicide belonging to the state.

Author's note – Tyrannicide was a 14 gun sloop.

In the month of December 1778 he entered into the service of the United States and received a commission as Sailing Master, dated in that month, on board the Frigate Queen of France, Captain Joseph Olney commander, and proceeded in her a cruise with a squadron commanded by Comodore John Hopkins which made several captures of the enemies ships & privateers that he after wards sailed upon another cruise in the same vessel under the command of Capt. John P Rathbone who percceded Capt. Olney and continued to serve as Sailing Master as aforesaid for the term of about twelve months when he was transferred to the Frigate Boston, Captain Samuel Tucker, on board which vessel he served two months until he was taken sick and removed into quarters. That from that time he was not called into the public service, there being as he believes no ships belonging to the United States in which he could be employed, and by permission of the Navy Board he took command of a Letter of Marque on board which he continued until peace took place.

Author's note – In some other correspondence in his pension claim folder it states that the Letter of Marque he commanded was the *Apollo.*

Henry Skinner's pension application was approved and he received $20.00 per year for his service under the 1818 Pension Act.

There is no mention of his date of death.

Cyprion Henry

Author's note – Cyprion Henry was not a sailor nor marine, but an agent for the United States. His deposition is very informative and was written to endorse a pension claim submitted by Samuel Johnson, which I will include after Mr. Henry's.

The deposition of Cyprion Henry of Providence in the State of Rhode Island who being engaged according to law testifies & says – That in the year 1779 this deponent was agent for the purpose of disposing of the Prizes which should be taken by the Queen of France, a vessel of War in the service of the United States & commanded by Joseph Olney – This deponent further testifies that said Vessel called the Queen of France was formerly an old French ship purchased by Silas Doans Esq, then an agent to the United States in France for the United States services – And after her arrival in this country she was manned, equipped & sailed on several cruises – First under command of John Hopkins with the Warren & Ranger in company – Afterwards she sailed on the second cruise in company with the Frigate Providence & Ranger under the command of Abraham Whipple – Lastly she sailed to Charleston S.C. where she was sunk, as he has always understood, when that city surrendered to the British, 1780, according to this deponents best recollection. – and this deponent further testifies that Samuel Johnson now of Walpole, Massachusetts whom he distinctly recollects, was Masters

Mate on board of said vessel during her first & second cruises & must have served on board her more than none months.

Samuel Johnson

I Samuel Johnson a citizen of the United States now residing in Walpole in the County of Norfolk and Commonwealth of Massachusetts on oath declare that I was born in Providence R.I. AD 1753. That about the middle of October AD 1778 I entered myself as a midshipmen on board the Continental Ship Queen of France commanded by Joseph Olney, was proceeded by Capt. John Peck Rothbone, William Thaxton was First Lieutenant, Joseph Cisey Second Lieutenant, Samuel Powell Third Lieutenant & Henry Skinner Sailing Master, all of whom except Skinner I believe have deceased. That soon after entering on board said ship I was promoted to Second Mate and thence to Chief Mate, and that I served in said capacity in the War of the Revolution upon the Continental establishment against the common enemy until September 1779, having a period of more than nine months until I was discharged.

Author's note – Samuel Johnson's pension application was approved and he received $8.00 per year for his service under the 1818 Pension Act.

There is no mention of his death.

USS Raleigh

The Raleigh was built after being authorized by the Continental Congress in 1775 and launched in 1776. She was classed as a 32-gun frigate and had a crew of around 180 men. She sailed under the American flag until being captured by the British in 1778.

Her captains are said to be Thomas Thompson & John Berry.

Pierce Murphy

I Pierce Murphy Lyman in the County of York & Commonwealth of Massachusetts aged sixty six & a citizen of said Commonwealth declare that about the month of June Anno Damini 1777 I entered on board the public armed ship of war called the Rawleigh of thirty two guns then commanded by Captain Thompson & lying in the Harbour at Portsmouth, New Hampshire – Engaged for one year as a sailor before the mast – We took two or three small prizes, but made very little from them. We went to France & returned to Boston where at the end of the year for which I engaged I was discharged.

Author's note – Pierce Murphy's pension application was approved and he received $8 per month for his service under the 1818 Pension Act.

There is no mention of his date of death.

Timothy Gleeson

I Timothy Gleeson of Loudon in the County of Rockingham & State of New Hampshire testify and declare –

Author's note – I am going to skip over Mr. Gleeson's land service and get right to his service on *Raleigh.*

My fourth enlistment was about the first of August 1777 under Captain Thomas Thompson Esq. Commander of the ship of war Raleigh – We sailed on a cruise from Piscataqua, August 1777.

Author's note – Piscataqua is in New Hampshire.

After taking some prizes we went into L'Oient in France and were detained twelve weeks in giving our ship a clean bottom. Then set out on our cruise to the southward & returned home in April or May following, on board of which ship I performed the duty of Steward during the term aforesaid to the general satisfaction of said Capt. & officers.

Author's note – Mr. Gleeson goes on to detail subsequent land service which I will omit.

He was granted a pension of $8 per month under the 1818 Pension Act.

He passed away on the 7th of February 1827.

John Frost

I John Frost of Perry in the County of Washington & State of Massachusetts, do testify & declare that some time in the month of November in the year one thousand seven hundred & seventy six at Portsmouth in the State of New Hampshire I entered on board the Frigate Rawleigh of thirty two guns but mounting thirty six, Thomas Thompson commander, as a midshipman – The said Frigate Rawleigh was a Continental vessel. I continued to do duty as a midshipman on board said Frigate until the month of May one thousand seven hundred & seventy eight when I was Honorably discharged at Boston in Massachusetts – The said Frigate lay for some months in Portsmouth after I entered on board of her waiting for her guns & other equipment when she sailed on a cruise – I was on board said Frigate in L'Orient in France when the first news of the capture of the army under Gen. Burgoyne was received by the Packet in France – The sloop of war Alfred of twenty guns, Hinman Commander, sailed from Portsmouth in company with the Rawleigh & continued in company almost all the time until after our being at L'Orient as above mentioned. The Rawleigh arrived in Boston in the spring of the year one thousand seven hundred & seventy eight & not long before I was discharged.

Author's note – John Frost's pension application was approved and he received $8 per month for his service under the 1818 Pension Act.

There is no mention of his date of death.

USS Randolph

The Randolph was authorized for construction in 1775 by the Continental Congress and launched in 1776. She was considered a 32-gun frigate and had a crew of around 320 men. She sailed under an American flag until she was sunk in 1778.

Here captains are said to be Nicholas Biddle.

Randolph – Nolan Van Powell

James Knight

State of Illinois, Edgar County.

On this twenty seventh day of September 1832 personally appeared in open court being a court of record to the Circuit Court in and for the said county of Edgar James Knight, a resident of the county of Edgar and state of Illinois aforesaid aged 82 years the 20th August last who being duly sworn according to law doth on his oath make the following declaration in order to obtain the benefit of the provision made by the act of Congress passed June 7 1832.

He states that he was born in the county of Philadelphia on the 20th of August 1750, this his age is given from the register in his father's family bible. That about the first day of July 1775 he enlisted in the service for twelve months in the county of Bedford and state of Pennsylvania under Robert Cluggage, captain in Magaw's battalion in the Pennsylvania regiment of rifle men commanded by Colonel Thomson, in a few days was marched to Carlisle in Cumberland county in said last named state, thence to Eastown is said state, thence to headquarters at Cambridge in the state of Massachusetts where we arrived about the last of August or first of September and encamped there till the 13th of March following. Then marched to the city of New York and after two weeks marched to Long Island and there got my discharge for my full term of 12 months service as aforesaid, which said discharge is lost. That about two weeks he again enlisted as a marine under Captain Shaw for 12 months and entered on board the Frigate Randolph of 32 guns, commanded by Captain Nicholas Biddle, William Barnes 1st Lieutenant, Thomas Douglass 2nd Lieutenant, and Joshua Fannin 3rd Lieutenant, and lay at Mud Island below Philadelphia until the 5th of February 1777 when said Frigate put to sea on a cruising voyage being about six weeks out put into Charleston in the state of South Carolina.

Author's note – The Randolph was commissioned in July 1776, so James Knight was on it commissioning crew. The cruise he mentions is her maiden voyage which escorted American merchant vessels out to sea. As she made her way to Charleston fever broke out aboard the vessel and a great number of the crew succumbed to the illness and were buried at sea.[2]

And while here his said second term of 12 months enlistment expired, but the captain would not give him a discharge, saying we must take another cruise and try to do something as we had done nothing yet. And there we lay until August in which time our main mast was twice struck with lightening and later repaired and the last time a conductor was placed leading from the mast head into the water. Sometime in August aforesaid set sail for another cruise and in 3 or 4 days fell in with a British ship of 20 guns, one of 12 guns, two brigs and a sloop of 8 guns all belonging to the British.

Author's note – The 20-gun vessel was the *True Briton*, laden with rum. The 8-gun sloop was the *Severn,* and laden with rum, sugar, ginger, and logwood. The two brigs were the *Charming Peggy* and the *L'Assomption*, both laden with salt. The 12-gun vessel is unknown.

With which said vessels we had an engagement and captured four of them, the sloop escaping. We brought them into Charleston, one Brig was discovered to belong to the French and was given up to them, the other 3 and cargo was sold under the prize law, the proceeds of which sales amounted to £658.15 to each private or soldier. The Randolph lay in Charleston till 15 December following when this applicant got his discharge, which is lost, and in a few days started home to Bedford County in the state of Pennsylvania where he arrived sometime in the month of March 1778. After 2 or 3 weeks he again enlisted for 9 months under Thomas Cluggage, captain, in Major Robert Cluggage's battalion. Marched to Sinking Spring Valley where we built a fort called the Lead Mine Fort and continued in service there till his said last 9 months term of service expired. He then continued in service as a volunteer for 12 months longer and served the whole of his last mentioned volunteer service and left the service in the year 1779.

Author's note – James Knight's pension application was approved and he received $80.00 per year under the 1832 Pension Act.

He passed away the 23rd of February 1838.

James Knight got off the *Randolph* just before she was lost on the 7th of March 1778 when she went up against the much bigger 64-gun HMS Yarmouth. The *Randolph's* magazine was hit and she exploded killing 311 of her crew. Only 4 survived the blast.[2]

John McPherson

7th of February 1820

I was wounded on the 10th day of September in the year of our Lord one thousand seven hundred & seventy seven on board the Randolph Frigate commanded by Nicholas Biddle Esqr. In action with a British ship of war called the True Briton commanded by Sir Thomas Venture whom we captured with three others at the same time & carried into Charleston South Carolina. I was there discharged in consequence of my wounds & granted a small pension of 20 shillings per month by the court of North County I think in the year 1785 which pittance I have drawn ever since, and it is necessary by a late law of Congress amongst other things, it is necessary to obtain a certified copy of the Certificate from the Secretary of War. I therefore presume on your goodness to obtain the same for me and send it to the Northumberland post office.

Author's note – John McPherson's pension application was approved and he received 20 pence for being wounded under the 7th of June 1785 Pension Act. At some point that amount became $2.66 per month, and in 1816 it was increased to $4.26 per month.

The pension a claim you just read appears to be for a claim under the 1818 Pension Act, but this cannot be confirmed from his folder nor if it was granted.

He passed away the 9th of August 1827

USS Ranger

The *Ranger* was commissioned in 1777 as a sloop. She was armed with 18 guns and manned with a crew of around 140 men. She sailed under the American flag until she was capture by the British in 1780, and then served in their fleet.

Here captains are said to have been John Paul Jones & Thomas Simpson.

Ranger engaging HMS Drake – Arthur N. Disney Sr.

Peter Masuere

I Peter Masuere formerly of Portsmouth in the State of New Hampshire, now resident citizen in the town of Perry, County of Coos and State aforesaid, on my oath testify and declare, that I have served in the Revolutionary War in the land and naval services of the United States on the Continental establishment more than nine months in the manner following that is to say, in the year AD 1775 I enlisted as a private soldier into Captain Elihue Dearing's company of Artillery, attached to Col. Gilmann's Regiment, stationed at New Castle in the state where I served about four months, when I was regularly discharged by commanding officer of said regiment. Some time after I was discharged from said service I entered on board the United States Frigate called the Rolla, sailed said vessel the first voyage she made as quartermaster, when we went to France from thence to the coast of Africa and then returned home.

I think the time I served on board said vessel on this cruise was about eight months when I was discharged from said Rolla was commanded by Captain Thomson –

Author's note – When he says *Rolla,* he means *Raleigh.*

I then entered as quartermaster on board the United States ship called the Ranger, a twenty gun ship, and sailed in said ship the second voyage she made out of Portsmouth, New Hampshire, Captain Thomas Simson commander. And on this cruise the Ranger in company with the Warren and Queen of France captured what was called the Georgia fleet, when we returned to said Portsmouth I was on said ship about three months when I was discharged. I then entered the second time on board the said Ranger as Captains Coxswain of the Barge under the command of the said Thomas Simson, we sailed from Portsmouth to Boston and joined Commodore Whipple of the Providence, and Captain Rathmore of the Queen of France when we proceeded on our cruise to the eastward, fell in with and took a part of the Jamaican fleet so called. I served in this last service about three or four months when I was discharged.

Author's note – Peter Masuere's pension application was approved and he received $8 per month for his service under the 1818 Pension Act.

There is no mention of his date of death.

Mark Staples

Mark Staples of New Market in the County of Rockingham and State of New Hampshire late a Seaman on board the United States Sloop of War Ranger of 18 guns, commanded by John Paul Jones Esq. In the Naval Service of the United States in the War of the Revolution, dulauth maketh oath & saith, that in the month of August or September 1777 or about that time at Portsmouth in said state he entered on board said sloop of war as aforesaid, commanded as aforesaid & in the service of aforesaid in said war, she lying in the harbour of Portsmouth, to serve the United States in said Sloop of Was asforsaid for one year – That he continued in said service on board until she sailed from Portsmouth on the first of November in said year for France & during her passage to France & the cruise she made from thence to the Irish sea & back to Brest to the United States under the command of Captain Simpson and arrived at said Portsmouth in October 1778.

Author's note – John Paul Jones left at Brest to take command of the *Bonhomme Richard,* leaving command of Ranger to Thomas Simpson.

His pension application was approved and he received $8 per month for his service under the 1818 Pension Act.

He passed away on the 3rd of April 1830.

William Blunt

Pensacola, West Florida, County of Escambia

On this Ninth day of June 1823 personally appeared in open court being a court of record for the District of West Florida William Blunt, resident in said county aged sixty years who being first duly sworn according to law doth on his oath make the following declaration in order to obtain the provision made by the acts of the 18th of March 1818, the 1st of May 1820, and the 1st of March 1823.

That he the said William Blunt shipped in the continental navy the year 1777. That the Rendevous being then supplied for the Frigate Raleigh then built at Portsmouth. The continental ship of war Ranger arrived from France called for him on board said ship on an expedition in company with the Boston and Charleston Frigates and Queen of France sloop of war for protection of our southern section, formed and sailed from Boston some time in November 1779 and by annoying the enemy embarking from New York & southerly, captured one gun boat and three transports on their passage to Charleston.

Author's note – Mr. Blunt had two written depositions in his claim and I am using the more detailed one. But he did omit in this one that Thomas Simpson was the captain of *Ranger* during this time.

Was then ordered with the ship on cruising ground where they captured 6 or 7 more of the enemies transports and sent them to Charleston – were driven into Charleston by the enemies two fifty gun ships and one seventy four commanded by Admiral Arbuthnot, when commenced the siege of Charleston by the enemy; and the surrender of Charleston May 12th 1780, our fate after 27 days cannonade. I arrived home again at my native place (Portsmouth N.H) after very suffering, being after some time redeemed. Our navy then dismissed, applied to the Department to ship on board the new ship Royal Louis 20 guns. Commenced on a cruise on the banks to annoy the Quebec and Halifax corn fleet, captured 3 prizes and was taken by a fifty gun ship, their convoy, and sent on board Proteus prison ship, was detained nine months a prisoner during which I suffered much. Was again redeemed and returned home; Shipped again in the St. Eastacia 16 guns, commenced cruising and capturing until taken by the enemy after hard fighting and left at the West Indies in prison.

Author's note – William Blunt's pension application was approved and he received $8 per month for his service under the 1818 Pension Act.

He passed away on the 1st of October 1831.

John Wheeler

I John Wheeler of Welton in the County of Kennebunk and District of Maine, aged sixty eight years, a resident citizen of the United States desiring to obtain the provision made by the law of Congress entitled "An Act to provide for certain persons engaged in the land and naval service of the United States in the revolutionary war," and to be placed on the pension list for said district of Maine, do on oath testify and say;

That at Portsmouth New Hampshire, about the month of September 1777, I shipped as quartermaster on the continental ship, or sloop of war Ranger on the continental establishment, Paul Jones Commander, in the war of the Revolution. Before we sailed information evolved of the capture of Burgoyne and we were detained to take dispatches of the intelligence to France. I was put on a prize brig, which we captured on our passage, and sent with her into Bordeaux, whence I joined the Ranger at L'Orient. We then cruised in the Bay of Biscay, afterward sailed with a French Frigate which left us near the Isle of Man. We captured a ship having on board furniture of the Lord Lieutenant of Ireland, and sent her into Brest. We attempted to cut out of Carrick – Fergus the British ship Drake of twenty guns, but failed by means of wind. We arrived in the Channel, sunk some vessels and took a light brig and carried into France, having set fire to the vessel in White Haven Harbour. At a subsequent day, before the taking of the brig last mentioned, the Drake came out and after a hard fight we took her and carried her into Brest. We Returned to Portsmouth and were verbally dismissed with honour, having served about fifteen months.

Author's note – John Wheeler's pension application was approved and he received $8 per month for his service under the 1818 Pension Act.

There is no mention of his date of death.

Danial Wise

United States of America, District of Maine,

On this twenty fourth day of July AD 1832 personally appeared before me the said Ashur Ware, Judge of the United States District for said District of Maine, Daniel Wise, a resident of Kennebunk in said District of Maine aged seventy one years – who being first duly sworn according to law, doth on his oath make the following declaration in order to obtain the benefit of the Act of Congress passed June7th 1832.

That he enrolled in the service of the United States and served as herein stated-

I entered & was a volunteer in a company of Militia in Portsmouth in the State of New Hampshire, commanded by Captain Hodsdon of Dover in the State of New Hampshire. I cannot recollect the year I enlisted in, AD 1776, 1777, OR 1778, the company was stationed New Castle or Great Island so called, on the south side of Portsmouth Harbour in said State of New Hampshire – the company built a Fort at Jerrell's Point, south of the light house. I served in said company from four to six months. I cannot recollect the time of service – at the expiration of this service I received at Great Island aforesaid a written discharge, which discharge is lost. I went immediately to sea and my wages were never paid. I think I was to have had forty shillings per month – I do not recollect the names of any other officers. The company passed muster at Portsmouth aforesaid I believe the reviewing officers name was Gaines, No other company was stationed at that place, Great Island during that time.

About the first of March seventeen hundred & seventy nine I enlisted or shipped as seaman at said Portsmouth on board the United States ship Ranger, Captain Thomas Simpson was commander and Elijah Hall First Lieutenant, both since dead. I believe my wages was eight dollars per month, the ship sailed from Portsmouth to Boston Harbour in the State of Massachusetts, from which place we sailed in company with the United States Frigate Providence and sloop of war Queen of France. While on the cruise we took ten Prizes, eight of which arrived in Boston aforesaid, the Prizes were taken from the Jamaican Fleet (so called). The Ranger returned to said Boston the last of August AD 1779.

I contained in the service and on board said ship Ranger until January AD 1780, when the Ranger then on a cruise in company with two other United States vessels took a Prize, the British Brig Dolphin – I was ordered on board said Prize – Within a month the Dolphin was taken by a British Privateer and I was stripped of all my clothing except my shirt & pantaloons. I was carried a prisoner to New Providence sometimes called Nassau and put aboard a prison ship. I ran away from the prison ship and arrived home the last of October AD 1780. I returned home as soon as possible. I had to obtain passage from New Providence to St. Lustecher & from hence home. It was between nineteen & twenty months from the time I shipped until my return. Soon after my return I shipped in Portsmouth aforesaid on board a Letter of Marque, the ship Alexander, then commanded by same Captain Simpson above mentioned. I went to France in said ship. On our voyage out we had an engagement with & took a British vessel of fourteen guns which was sent to and arrive at Boston aforesaid. I served on board said ship Alexander about eleven months, when she returned to Boston, where I was discharged.

After my return in the aforementioned ship, I shipped at said Portsmouth on board an American Brig, which I can not recollect. The Brig was commanded by Captain Turner. We sailed from Portsmouth to Martinico. On our return voyage we captured a British ship of fourteen guns and carried her into Boston. I served on board said vessel five or six months & was discharged at Boston aforesaid, from which time I was at sea on board privateers most of the time until Peace.

Author's note – Daniel Wise's pension application was approved and he received $92.66 per year for 19 months and 5 days of service under the 1832 Pension Act.

There is no mention of his date of death.

William Stacey

I William Stacy of York in the County of York and State of Massachusetts testify & say that some time in the month of April AD 1776 I enlisted as a private soldier into the service of the United States in the Revolutionary War, in the company commanded by Captain Tobias Fernald, James Donnell being the First Lieutenant of said company. The Regiment in which I served was commanded by Colonel Finney or Phinney – I enlisted in the twelve month service which twelve months expired some time in December 1776, at which time I was honorably discharged at Albany in the State of New York having served about eight months. I also served about seven months at Fort Sullivan on Piscataqua River under Captain Eliphalet Danials and from February 1777 to September 1777. I further testify and say that some time in the month of September 1777 I enlisted as a marine on board the United States Sloop of War Ranger, John Paul Jones commander. We sailed in the month of October from Portsmouth New Hampshire for France and from France on a cruise in the English Channel, we landed at White Haven in the North of England during our said cruise and took the Fort at that place, spiked the cannon, and burnt the shipping in that port. Afterwards landed in Scotland and paid a visit to Lord Mansfield, afterwards cruised on the Irish Coast and there took the Sloop of War Drake lying in the Harbour of Belfast and returned to Brest in France with our Prize.

And from France returned to Portsmouth New Hampshire in the month of October 1778 making a cruise of about a year and where I was honorably discharged. We left Capt. Jones in France and Captain Thomas Simpson came home in the ship from France as commander.

I further testify and say that in the month of April 1779 I shipped on board the same sloop of war Ranger, the commanded by the said Captain Thomas Simpson, at Portsmouth New Hampshire as an ordinary seaman, in which capacity I served about six months – We sailed from said Portsmouth to Boston where joined the squadron under the command of Commadore Whipple, and sailed from Boston in company with the Frigate Providence commanded by said Commadore Whipple and the sloop of war Queen of France. We cruised on the American Coast, the Bank of Newfoundland and as far as the Channel of England and the squadron took ten sail from the Jamaican Fleet during this cruise; and we returned to Boston in the fall of the year 1779 where I was again honorably discharged.

Author's note – William Stacey's pension application was approved and he received $8.00 per month for 12 months of service under the 1818 Pension Act.

There is no mention of his date of death.

USS Resistance

The *Resistance* was a brigantine mounted with 10 guns launched in 1777. She sailed under the American flag until being captured by the British in 1778.

Her captains are said to be Samuel Chew.

Jonathan Averel

State of New York, Herkimer County

On this Eleventh day of May 1818 before me the subscriber one of the Judges of the Court of Common Pleas of the County of Herkimer aforesaid personally appeared Jonathan Averel aged sixty two years, resident in the town of Warren in the County aforesaid, who being by me first duly sworn according to Law doth on his oath make the following declaration in order to obtain the provision made by the late act of Congress entitled "An Act to provide for certain persons engaged in the Land & Naval service of the United States in the Revolutionary War." That the said Jonathan enlisted as a musician in the State of Connecticut in Captain Waterman Clefts company in

Regiment commanded by Colonial Holden Parsons. And about the month of April in the year 1775 and continued in said company about eight or nine months, stationed at Roxbury, and in the year 1776 enlisted in the company commanded by Captain Oliver Coit in Colonel John Ely's Regiment as a musician & continued in the said company for the term of eight or nine months during which time he was in the Battle of White Plains in the year 1776. And sometime in the month of June in the year 1777 received the appointment of Second Surgeon on board the Brig Resistance, commanded by Captain Samuel Chew, and on the 27th of January 1778 received the appointment of full Surgeon on board the same vessel, and continued on board said vessel about twelve months in actual service, during which time he was in three severe actions, in one of which Captain Chew was killed, and sometime in the month of December 1778 he received the appointment of Second Surgeon aboard the Frigate Providence, Frigate commanded by Abraham Whipple, and continued on board the same vessel, being in actual service for about eight or nine months, and sometime in the month of May received the appointment of Second Surgeon on board the Frigate Alliance, but I never did any duty on board of same vessel, and in the year 1781 I received the appointment of First Surgeon on board the Brig Wexford, commanded by John Peck Rathbon, and continued on board of same vessel about two months when the vessel was captured & taken by the British and carried in to Ireland, together with the crew as prisoners, in which place he lost all his papers, in which place he remained about three *months, and then made his escape to Portugal and from thence returned to his native country where he at first enlisted.*

Author's note – Jonathan Averel's pension application was approved and he received $480.00 per year for his service under the 1832 Pension Act.

He passed away the 16th of August 1835.

Daniel Brown

New York Mayors Court

Present the Honorable Judge Peter A Jay, recording, of the City of New York, for be it remembered that on the twenty third day of September in the year of our Lord one thousand eight hundred and twenty three personally appeared in the Court of Common Pleas called the Mayors Court being a court of record for the City of and County of New York, according to the charity of the said city and the laws of the State of New York, Daniel Brown, aged sixty seven years, resident in the City of New York who being first duly sworn according to Law doth on his oath make the following declaration in order to obtain the provisions made by the acts of Congress of the 18th of March 1818, and the first of May 1820, that he the said Daniel Brown shipped on board the Brig Resistance at New

London in the year 1777 mounting 16 guns and commanded by Captain Samuel Chew as a Midshipman bound on a cruise to the Windward Islands, in the line of the State of Connecticut, the Connecticut Continental Establishment, that he continued to serve in the said vessel until the year 1778 when he was honorably discharged from the said service in Boston by John Deshow, agent of the Navy department of that place.

They proceeded from Connecticut to Martinique, and previous to their arrival they took four prizes and sent three of them into Martinique; after having touched there with their fourth prize they fell in with and had an engagement with the Cataurit Paches mounting 22 guns when Captain Chew was killed and the brig hauled off and the Sailing Master took command, the first lieutenant having been wounded & the 2nd lieutenant lay sick with a fever. They went into Martinique and was ordered home from there by the American Consul and came home to Boston where he was discharged as aforesaid.

Author's note – Daniel Brown's pension application was denied. The War Department stated that the *Resistance* did not belong to the navy of the United States. There was also no mention of his death.

William Leeds

Author's note – This is an unusually written declaration and appears to be notes taken from the original pension claim. But it is in his pension folder.

Leeds, William

His name appears on a list of applicants for invalid pension retired by the District Court for the District of Connecticut, submitted to the House of Representatives by the Secretary of War on December 31, 1794, and printed in the American State Papers, class 9, page 142.

Rank: Lieutenant

Armed Brig Resistance, Captain Chew

Disability: Wounded by a musket ball in his shoulder during an engagement with a British Letter of Marque in which action the command desolved upon him, Captain Chew having been killed

When and where disabled: Winter 1777, 1778

Resistance: New London

To what pension entitled: One half

Remarks: Entered on board the Resistance July 5, 1777 and wounded the 4th of March 1778, as per roll, and his letter to the Marine Committee, dated March 20, 1778. Evidence incomplete, only one witness to prove his being wounded in the line of his duty in actual service. This witness deposeth that he does not know any person now living who was on board the brig at the time of the engagement excepting the claimant. 2nd, the report of the examining physician is not upon oath. 3rd, the evidence of the three freeholders testify his disability, mode of life, since the year 1791 only, he having been left at Martinique after the engagement, where he remained in consequence of his wound till that year. Other evidence is produced to support his claim, which is taken before Justices of the Peace and not by the District Judge, or commissioners as required by law.

Author's note – William Leeds was placed on the Pension Roll on the 4th of September 1795.

He passed away in 1804.

USS Saratoga

The *Saratoga* was a sloop mounted with 18 guns, manned with around 90 men, and launched in 1780. She sailed under the American flag until she was lost at sea in 1781, presumably in a storm.

Her captains are said to be John Young.

Elkanah Cook

Commonwealth of Massachusetts, County of Plymouth

On this twenty ninth day of August 1832 personally appeared in open court before Honorable Wilkes Wood, Judge, the Court of Probate, now sitting at Scituate within and for said County of Plymouth, Elkanah Cook, a resident of Kingston in said County of Plymouth, and commonwealth of Massachusetts, aged seventy three years, who being first duly sworn according to law doth on his oath, make the following declaration, in order to obtain the benefit of the act of Congress passed June 7th, 1832. He hereby relinquishes every claim whatever to a pension or annuity except the present, and declares that his *name is not on the pension roll of the agency of any State. That is to say, that he entered the service of the United States under the following named officers, and served herein stated.*

Author's note – I am going to skip over his initial land service and jump right into his sea service.

In the year 1779 he sailed from Baltimore in the schooner Flying Fish, Capt. Sacket, Master, bound to St. Eustacia. As soon as clear of the Capes was taken by an English Privateer out of New York. In four days after was retaken by the United States 20 Gun Ship Saratoga, Capt. Young commander, from Philadelphia bound on a cruise off Charleston and the West Indies. Capt. Young told him if he would ship for service he should have the same wages & prize money as those who shipped in Philadelphia. He accordingly shipped and chose his Capt. as his agent. In five or six weeks fell in with and English Letter of Marque ship of 20 guns. Engaged & took her. She was from Charleston bound to Liverpool with valuable cargo, in the action the English lost 11 men killed & nearly twice that number wounded. In four or five days fell in with & captured a Scottish Brig of eight guns loaded with wine. The Capt. Put me, said Elkanah Cook & others on board as Prize Crew & arrived at Philadelphia. He heard of the arrival of the Saratoga at Cape Fransna with her prize, which was the last he heard of her. Some prisoners arrived from Charleston & brought news that such a ship was lost in a squall in company of a ship that arrived there. He remained in Philadelphia til the agents advised him to go a short voyage. Got a voyage to the West Indies. Was taken by the English. Sent to England first on board a man of war ship & there remained until the end of the Revolutionary War. About three years after went to Philadelphia and was told that everything pertaining to his cruise in the Saratoga was settled two years previous, and nothing there remained in the hands of the agents. Thus he lost <u>*six & half months*</u> *wages and his share of Prize money.*

Author's note – Elkanah Cook's pension application was approved and he received $27.33 per year for his service under the 1832 Pension Act.

There is no mention of his date of death

USS Trumbull

The *Trumbull* was a 32-gun frigate purchased by the Continental Congress and commissioned in 1776. She was manned with around 200 men and sailed under an American flag until being captured in 1781.

Her captains are said to be Dudley Saltonstall, Elisha Hinman & James Nicholson

Silvanus Mather

In order to obtain the benefit of the Act of Congress passed June 7 1832.

State of New York, County of Otsego

On the seventeenth day of October 1832 personally appeared in open court before the Judges of the Court of Common Pleas in and for the County of Otsego in the State of New York now sitting Silvanus Mather a resident of the town of New Lisbon in the County of Otsego and State of New York, aged seventy-one years who being first duly sworn according to law, doth on his oath make the following declaration in order to obtain the benefit of the act of Congress passed June 7th 1832. That he entered the service of the United States under the following named officers and served as herein stated.

That in the time of the Revolutionary War he resided in the town of Lyme New London County in the State of Connecticut. That about the first of May 1778 he enlisted under Captain Dudley Saltenstall commander of the ship Trumbull for no distinctive term, but to serve as a guard, to guard said ship while she lay blockaded at the mouth of Connecticut River. The 1st Lieutenant's name was Marlborough of Fairfield, 2nd Lieutenant's name was John White of Middleton were on board and he continued in said service the term of sixteen months, till about the last of August 1777, when he was discharged. The ship was built up the river at Chatham and manned and equipped in order to go to sea. But could not get over the bar at the mouth of the river on account of her draft of water. While she was detained there the British sent out a Frigate to Blockade and take her if she could get over the bar. The Captain pleading that he could not get to sea with his ship, removed the seaman & marines and enlisted a guard of about 20 men to guard the ship, of which he was one. While he was on board said ship Capt. Saltenstall left Trumbull in the command of Capt. Hinman and went on board of some other vessel and was Commodore of the fleet, had a Battle with the enemy and got his fleet badly tore to pieces as this deponent understood at that time. The British abandoned the blockade of the Trumbull and she was finally got out of the river to New London and fitted out to sea.

After he left her, and as he heard, she had a Battle with the enemy and got crippled, but escaped to Boston, was again fitted out and taken by the enemy and carried to New York and condemned and made a prison ship of her. He did not receive a written discharge.

Author's note – Silvanus Mather's pension application was approved and he received $50.00 per year for his service under the 1832 Pension Act.

He passed away the 1st of February 1835.

Bordon Wilcox

On this sixteenth day of October in the year of our Lord one thousand eight hundred and thirty two, personally appeared in open court before the Judges of the Court of Common Pleas, now sitting in the town of Batavia in and for the County of Genesee and State of New York, Bordon Wilcox a resident of said town of Batavia in the said County of Genesee and said State of New York, aged seventy one years on the third day of February 1832, who being first duly sworn according to law, doth, on his oath make the following declaration in order to obtain the benefit of the Act of Congress passed June 7, 1832.

Author's note – Mr. Wilcox's pension application starts with a lengthy land service, so I will jump in at the beginning of his sea service.

The month of April 1780 when he enlisted for eight months as a mariner on board of the United States Frigate Trumbull of 32 guns, that he enlisted at New London in Connecticut and received eight dollars per month – that the ship was commanded by Capt. James Nichoson, that the first lieutenant was Maltby and that there was Lieutenant White and Starr. That soon after he entered the ship she put to sea and cruised off the south of the Island of Bermuda and that on the first day of June 1780, they had an engagement with a British Ship by the name of Watt from Liverpool loaded with money and clothing for the troops at New York – that the battle last about five glasses and ten minutes – a drawn battle.

That the mainmast, amizenmast and fortop mast of the Trumbull were cut away and 48 men killed. That Lieu. Starr was killed in the engagement by a chain shot which took off his head. That after the battle we got up some masts and ran into Boston and there rigged out and went to sea again and carried continental stores to Philadelphia for other ships which were rigging out there. Then left Philadelphia and went on another cruise off the Capes of Virginia and were chased off by two British Frigates, when they returned to Philadelphia and he was discharged –

That he went on board a Privateer called the Morning Starr and cruised off Charleston when a British Frigate gave chase and took ship and carried her into Charleston where he was detained as a prisoner eleven months. That after he obtained his liberty he went on several other privateering voyages.

Author's note – Borden Wilcox's pension application was approved and he received $87.61 per year for his service as a seaman for 8 months and a private for 16 months under the 1832 Pension Act.

He passed away the 22nd of April 1848.

Richard Law

I Richard Law of New London, State of Connecticut under oath, make this declaration, in order to obtain the benefit of the Act of Congress passed the 7th day of June 1832. I am now a resident of New London aforesaid which is the place of my birth, and in the seventieth year of my age.

That I was appointed and entered as a midshipman on board the United States Frigate Trumbull in the early part of the year 1779, at that time said ship was laying in the River Thames above New London and commanded by Elisha Hinman Esq who was superseded by Capt. James Nicholson in the same year before the ship left for Boston – That I continued in said Ship Trumbull as midshipman until the time of her capture by the British in the month of June or July A.D. 1781 when I was with the rest of the officers and crew of said Ship Trumbull carried into the Port of New York and confined on board the Jersey Prison Ship from which situation I was paroled in the month of September 1781, when I returned in a coastal to New London in Connecticut, and my exchange was not effective until some time after the prison date & I do not recollect. I was in actual service for more than two years as aforesaid midshipman.

Author's note – Richard Law's pension application was approved and he received $144.00 per month for his 2 years' service as a midshipman under the 1832 Pension Act. I did double check this amount and payment schedule, since most 1832 pensions were paid annually.

There is no mention of his death.

John Trevett

On the twentieth day of May AD. 1820 personally appeared in open Court being a Court of Record for the County of Newport, State of Rhode Island & Providence Plantation, United States of America, John Trevett, in the seventy second year of his age, a native Citizen of and resident in said County aforesaid, who being first duly sworn according to law doth on his oath declare that he served through the War of the revolution to wit that he entered on board of the Ship Collumbus in Nov. 1775 as a midshipman, Abraham Whipple Esq commander, and in Feb. following he was appointed a commission as First Lieutenant of Marines under the command of Eseck Hopkins, and after a cruise I went on board the Ship Providence, Hoisted Hacker Com. – He then being under the command of John Paul Jones – I then took command of the Marines on board said ship and we sailed on a cruise, and the beginning of November following, we then being to the eastward of Halifax, we captured several British Ships and one Snow, and got them all safe into New Bedford, one of the ships having Kings stores on board, and bound to Quebeck with ten thousand suits of soldiers clothing for Gen. Burguines Army. We saw fit to send them to General Washingtons Army – At this time Jonathan Pitcher took the command of the ship Providence, I continued the command of the Marines, and in March 1777 we sailed on another cruise to the Eastward, where we fell in with the Letter of Marque Brig called the Lucy, mounting twelve sixes, bound to Quebeck, we captured her, we then proceeded to New Bedford to repair. After this John Peck Rathbone took the command. I belonged to this vessel two years, and in 1780 I had orders to go to New London where I went on board the Continental Frigate Trumbull, commanded by James Nicholson Esq. We sailed from New London in May 1780. A few days after we captured an English Privateer and in cruising off Bermuda we fell in with a Ship called the Walter of thirty guns. This being the first of June, we had a smart engagement with her in which we had in officers & men, forty three in killed and wounded, all I got for this cruise was the loss of my right eye! And a wound in my right foot!

I was next on board of the ship Dean, Elisha Hinman Esq. We captured several Prizes and in one of them I was retaken and carried in to St. John Newfoundland, where I was detained two years and one month for my friendship to King George I arrived in Philadelphia in June 1782 after several afflictions and having my Constitution much injured and with one eye.

Author's note – John Trevett's pension application was approved and he received $20.00 per month for his service under the 1818 Pension Act.

There is no mention of his death.

David Phipps

Author's note – David Phipps declaration is found in Christian Hanson's pension claim folder where he is vouching for Mr. Hanson's service.

State of Connecticut, New Haven County

On this 4th day of November 1818 before me Simon Baldwin, Justice of the Peace in aforesaid county, personally appeared Capt. David Phipps of New Haven in said county now in commission as a Sailing Master in the service of the United States to me personally known, a gentlemen whose character as to credibility is exceptionably fair & being first duly sworn deposeth, that on the 1oth of Oct. 1776 he was commissioned Second Lieutenant of the continental Frigate Trumbull of 30 guns, Dudley Saltonstall commander, a new ship then lying in Connecticut River. That he continued to serve on board said ship until 15 Jan 1778. Unable to get her to sea.

That in May 1777 Christian Hanson enlisted to serve on board said ship for the term of one year & served as Cabin or Captain's Cook & continued to serve on board until the deponent left said ship to go on board the Warren, then lying at Providence – That he then left Hanson on board the Trumbull & has no doubt he continued to serve on board till the term of his enlistment , one year, expired & that the Trumbull was then blockaded & obstructed, & with much exertion did not & could not get to sea during that year & for a long time after.

Author's note – Christian Hanson's pension application was approved and he received $8.00 per month for his service under the 1818 Pension Act.

There is no mention of his death

USS Warren

Schooner

There were two vessels named *Warren* during this period, a Schooner & Frigate. The Schooner *Warren* was a merchant vessel purchased by the Continental Congress in 1775. She carried a crew of about 50 men and armed with 4 cannons and 10 swivels. She sailed under the American flag until being captured and sunk in 1776.

Her captains are said to be Winborn Adams & William Burke.

John Cushing

State of New Hampshire County of Hillsborough

On this 27th day of August 1832, personally appeared in open Court before the Judge of the Court of Probate in said county, now sitting John Cushing, a resident of Goffstown in the County of Hillsborough and State of N.H. aged 82 years, who being first duly sworn according to law, doth on his oath make the following named officers, and served herein stated- that is to say –

In October AD 1775 he enlisted aboard of the privateer Brig Washington & sail'd out of Beverly, Massachusetts, Elias Smith commander, Benjamin Leavett 1st Lieutenant, Moses Leavett, Sailing master, & took five prizes – one was retaken and four brought into Salem. Mass. Was out three months – In November AD 1776 entered aboard of the privateer schooner Warren of Salem, Mass. And after being out about three weeks was taken by the British Ship Thom, Letter of Marque from Liverpool. Was carried into that place & kept a prisoner till May following when he was sent to Mill Prison and confined there two years & seven months – He was then exchanged and wen to St. Marlow in France & then to L'Orient, at which place he enlisted for twelve months in the service of the United States on board of the ship Carolina, from South Carolina, commadore Alexander Gillon commander, John Joiner Captain, Nicholas Bartlett First Lieutenant. He enlisted in May or June AD 1780 as a Surgeons Mate at Twenty Dollars per month – From L'Orient he went to Amsterdam and remained in the service with said ship nineteen months & twenty one days, and then left the ship on her return to America at Havana, from whence he returned to Salem in a Brig commanded by Captain Waters.

Author's note – John Cushing's pension application was approved and he received $393.33 per year for 19 months and 20 days of service under the 1832 Pension Act.

There is no mention of his date of death.

Thomas Doliber

Commonwealth of Massachusetts

Be it remembered, that Thomas Doliber aged 62 of Marblehead in the County of Essex in the Commonwealth - mariner – aforesaid, doth make the following Declaration, to, wit,

That he enlisted into the service of the United States in the month of May, anno domini 1775 in Captain Merret's company, Col. John Glover's Regiment of infantry in the Massachusetts Line. Marched from Marblehead to Cambridge where we served about four months. Was then ordered by General Washington on board the Continental ship Lee, John Manley commander to cruise against the enemy. That he continued the same service until his enlistment expired about the last of December 1775. That he never had any written discharge – A few days after he again enlisted in the service of the United States in Captain Enoch Putnam's company, Col. Hulchinson's Regiment of infantry in the Massachusetts Line. He marched from Marblehead to Winter Hill in Charlestown, Massachusetts. Continued in the vicinity of Boston until the next July following when he was ordered on board the Continental schooner Warren, William Burke Captain, to cruise against the enemy and continued on board until she blew up in an action with a British ship. That he was badly wounded and unfit for service. About the first of November 1776 he was discharged by general Ward and by some accident his discharge has been lost.

Author's note – Thomas Doliber's pension application was approved and he received $8.00 per month for 9 months of service as a private under the 1818 Pension Act. His sea service does not seem to have been credited.

There is no mention of his date of death.

USS Warren

Frigate

The Frigate *Warren* was contracted to be built by the Continental Congress in 1775 and launched in 1776. She was armed with 32 guns and manned with around 250 men. She sailed under the American flag until she was burnt in 1779 to prevent being captured.

Her captains are said to be John B. Hopkins & Dudley Saltonstall

Nehemiah Manson

State of Massachusetts, Plymouth County

On this twenty second day of August eighteen hundred and thirty two personally appeared before the Hon. Wilkes Wood Judge of the Court of Probate for the county of Plymouth now sitting at Hanover in said County, Nehemiah Manson, a

resident of Situate in said county and State of Massachusetts aged seventy one years, who being first duly sworn according to law, doth on his oath, make the following declaration in order to obtain the benefit of provision made by the Act of Congress passed June 7 – 1832.

Author's note – Mr. Manson's pension claim is long and covers a lot of land duty, so I will just enter his time aboard *Warren* – which is his only time at sea.

In May of the fore part of June A.D. 1779 I enlisted on board of the ship Warren – Commanded by Commodore Saltonstal belonging to the State of Massachusetts, then lying in Boston harbour which had just arrived from a cruise off New York, and proceeded immediately to Penobscot, in a few weeks after arriving, the squadron under command of Commadore Saltonstall were attacked by the British Fleet. When one vessel only, a Brigg called the Angelico escaped and went to sea, one ship went on shore, and the remainder of the fleet were driven up the river, and towards the close of the term for which I enlisted being three months the vessels were all burnt and destroyed and I returned to Boston by land.

Author's note – Nehemiah Manson's pension application was approved and he received $82.00 per year for 21 months as a private and 3 months as a seaman under the 1832 Pension Act.

He passed away the 13th of November 1832, not three months after giving his deposition of service.

Edmund Burr

District of Virginia, Stafford County, February 9th 1829

On this 9th day of February 1829 personally appeared in open court for the county of Stafford, Edmund Burr, aged sixty eight last September, formerly of the State of Connecticut, now resident of the County of Stafford & being duly sworn according to Law doth on his oath make the following declaration in order to obtain the provision made by the Act of Congress of the 18th of March 1818 & the first of May 1820. That the said Edmund Burr in June 1776 inlisted as a soldier in the State Line of his native State in the company of Capt. Elijah Abel attached to Col. Philip Bradley's Regiment and that he was one of the few left of that Regiment after Fort Washington was taken. Was in the retreat through the Jersey's, was dismissed from the Army near Trenton, time of service being expired.

In the spring of 1777 entered as a sailor on board the Frigate Trumbull lying in Connecticut River, Dudley Saltonstall Esq commander – Jonathon Malsbie First Lieutenant – The Trumbull not going out of the River that summer on account of the shoal water on the Bar & the enemy's ships lying in Garner's Bay – In the month of January 1778 was turned over to the Frigate Warren then lying at Providence Rhode Island commanded by John Hopkins Esq with about forty of my ship mates. David Phipps Second Lieutenant of the Trumbull went with us & was First Lieutenant of the Warren – The enemy at that time in possession of Rhode Island, we run the gauntlet by the enemy's ships & went to sea in the Warren. Took a prize near Bermuda in sight of the Island, the ship Neptune from Whitehaven (England) loaded with salt & dry goods bound to Philadelphia, then in possession of the enemy. Was put on board the prize ship John Brown, privateer, shaped our course for Boston, was retaken off Cape Ann by the Orpheus Frigate, Sir Charles Hudson commander, was compelled to do duty until taken down with the small pox the natural way, about sixty days after we were retaken arrived at Rhode Island where our prize master died with the small pox & some more of my ship mates. The few of us that survived the small pox were immured in the prison ship until exchanged & sent to Providence some time in May as well as I recollect – Thence I proceeded home to my Friends with a broken constitution & was under the care of a physician until the October following, my life being dispaired of my friends. The result of hardship & ill treatment by the enemy whose tender mercy at that time was cruelty – The whole time of service on board of ship about fifteen months.

Author's note – Edmund Burr's pension application was approved and he received $8.00 per month for 15 months of service under the 1832 Pension Act.

There is no mention of his death.

George Long

To the Honorable Lewis Cass, Secretary of War, Washington D.C.

A history of the service of George Long of Portsmouth New Hampshire, during the whole of the War of the Revolution after 1775-

I George Long of Portsmout, State of New Hampshire, Merchant of the age of seventy years & upward – Son of Col. Pierce Long who commanded the First New Hampshire Regiment – Testify and say –

Author's note – Mr. Long's declaration is lengthy and I am going to jump in on page four during the year 1778.

In the same year your deponent engaged in the sea service in a ship of 20 guns & 130 men & did go with Capt. Moses Brown, James McClure was First Lieut. (the same who had been adjutant of Col. Long's Regiment) In this ship we encountered a British ship – about one-hour & captured her; in this action we had 2 killed & 9 wounded; among the latter, Lt. McClure lost his leg –

Author's note – In another section of his deposition it describes the ship he was on was the *Gred Arnold*. The ship they fought was the *Adventure,* commanded by Capt. Bretulu.

In May 1779 I entered on board the Polly – John Palmer commander, 10 guns & 25 men – Our 1st Lt. was James Falls, 2nd Lt. Burns – your deponent Masters Mate. We sailed in company with three other ships, the Sally – Capt. Holmes 18 guns & 60 men, Minerva – Capt. Grimes 10 guns & 40 men, Cadwallader – 10 guns & 30 men. After being at sea two days we encountered a British Ship named Blaize Castle – Capt. Shepherd of 22 guns & 130 men from Halifax. The ship was engaged by whole fleet, alternately, for two & half hours, when she struck her colours. We had one killed & Capt. wounded. The others had more or less kill'd & wounded. The whole of us not so many as the enemy. They had laying on her deck (when your deponent went on board the enemy, immediately after striking to us) 13 dead men & they told us they had 30 -40 wounded – With our prize we all came to Boston to repair damages - which were considerable in sail & rigging. After refitting, and nearly ready for sea, an embargo was laid on all vessels in port, for the reason that an armament was fitting out by the State of Massachusetts, under sanction of Congress, called the Penobscot Expedition; for the purpose of recapturing a fort on Penobscot River, then an internal part of Massachusetts, which had been in possession of the British some time. There were assembled in Boston Harbour about 20 Ships & Brigs and other armed vessels for this expedition. Among which were TWO, belonging to the U. States, the largest was the U.S Frigate Warren, Dudley Saltonstall Esq Commodore of the fleet, 36 guns. The other the sloop Providence – 10 guns. A few days before this directive of the fleet for Penobscot, say July, a boat with several officers came on board our vessel, the headman's name was Dane or Dana, as I heard him addressed. He ordered our Capt. to muster all his crew fourth with – which was done. We were then told by this gentleman that we must all go on board the U.S. Frigate Warren – Commodore Saltonstall, she wanting seaman – to that we one and all objected. We were then told if we did not go freely, we would be compelled so to do. A guard was left on board our vessel as well as on board every one of our consorts, as I understood. The next day we were transferr'd on board the U.S Frigate Warren. Your deponent asked the officers from shore if we were impressed, the answer was you may call it what you please. Shortly after being on board the Warren Frigate we the impressed men were mustered, when a Roll or Shipping Papers was revealed to each of us to sign, which your deponent not only refused to do but advised all his associates NOT to sign. We were told by a Lieutenant, if we did not sign the Ships Roll we would not receive any WAGES – which proved too true – I never received a cent. I observed to him, if we sign the article we should be considered as having voluntarily become part of the crew – which was not the case – as we were impressed men, we chose to remain as such. Being a Petty Officer on board our own ship, I acted as a midshipman on board the Warren. The fleet all left Boston in July, of the date I am not positive, in a few days arrived before Penobscot (called Baggadee) where much time was consumed in preparation for war, and nothing decisively done. Suffice to say a superior British fleet arrived from New York with 1500 men to reinforce the Garrison – the fleet was under the command of Sir George Collins. It proved sufficient to prevent the escape of the American fleet, which were all captured. What did not get run up the river was destroyed by Americans – Your deponent with many others were captured and sent to Halifax and there imprisoned, we were called rebels & was so treated. All but starved. *The treatment caused great mortality among the prisoners, that 2 from every 3 died from two crews belonging to Portsmouth in about 8 months. Your deponent was exchanged 25th of Dec 1779, arrived at Cape Ann Jan 1780 having been absent from Portsmouth eight months. Had been sick twice in prison with the prevailing fever, and when arrived, too sick to be exposed as I was compelled to be before I reached my fathers house, to which your deponent was immediately confined with the same fever. From which I was unable to be abroad until the May following, making one whole year. All this occurring by reason of being impressed from 1778 to May 1780, and then not able to do anything until the fall of that year.*

In 1781 January your deponent entered on board ship Hector – Thomas Manning Commander, 18 guns with 100 men. Few months we were captured by the British Frigate Virginia of 32 guns. Carried to Bermuda. Was soon released. Arrived at Portsmouth. Went next on board Sloop Fox. Same Captain Manning. Few weeks we captured a British ship, very valuable. One of which your deponent brought into port 1781. Next on board Cutter Greyhound, 8 guns – Samuel Stuey Esq commander. With this vessel we captured a Brig of equal force by boarding. She was taken with Brandy, wine, with a few bales of dry goods – She arrived safe, but we were captured by the Assurance, 44 gun ship, Sir Andrew Douglas commander. Carried to New York – where we suffered badly- After a few months were released. In 1782 I engaged myself as Lieutenant on board the Brig Scorpion of 12 guns – John Stroble commander. In this vessel I meet the peace of 1783.

Author's note – George Long's pension application was approved and he received $42.66 per year for his service under the 1832 Pension Act. After all that sea time, only six months was credited from his time as a midshipman aboard Warren.

There is no mention of his death.

Sea Stories

As I mentioned in the introduction, there were other vessels that fought for America's Independence besides those in the Continental Navy. I was not going to include these types of vessels in this book, but then I came across Oliver Johonnot's & George Pillsbury's pension applications, and changed my mind. Their declarations demonstrate the enormous effort provided by these wide range of armed vessels fighting for American Independence.

Oliver Johonnot

The undersigned represents that he is a native citizen of the United States, that he was born in Middletown Connecticut in the year 1760. That in Sept. 1776 at Boston Massachusetts he entered as a sea man the Naval service of his country (in the Revolutionary War) on board of an armed Brig of 16 guns called the Rising Empire (& by the seaman Rising States) James Thompson Esq Commander, commissioned by Congress to cruise against the English, , in the month of December of the same year we sailed in the said Brig on a Cruise, in two or three weeks after we fell in with & captured an English Ship from White Haven England (Jonsenby Capt.) bound to Jamacia, she was maned & ordered for Boston, but was recaptured & her prize crew carried to India & never returned till after the War was over. We then proceeded to cruise off the Western Islands, there captured an English Brig & Sloop, sent them to France, then to the Bay of Biscay, our cruising station, but was taken in the month of April by an English seventy four gun ship the Terrible, Commander Sir Richard Bickerton, and retained on board this ship from April to June, after her return to Spit Head, until a Prison could be prepared for us. In July we were landed from the Terrible and carried before & examined by the Lord of the Admiralty who declared our doom (as they said hanging) but thank God the capture of Burgoyne & his army attacked their conduct & treatment towards us when we received the happy news of Burgoyne's fate. Myself with twenty nine others was confined in a Dungeon for attempting to make our escape, we were put upon half the prisoners allowance, eight ounces bread, six beef & some water for 24 hours. We remained in prison (Fortune Prison near Portsmouth) until June when we were exchanged & sent to France, on our same month we found three ships of War belonging to the United States, namely the Bonne Homme Richard Commander John Paul Jones, Frigate Alliance Capt. Landais & ship Genl. Mifflin of 20 guns Capt. George W Babcock. In the month of August 1779 the exchanged prisoners were assigned by a United States Agent or Commissions among the above named ships, myself as a seaman on board the Mifflin, the three sailed nearly together on that cruise, the

Genl. Mifflin captured the British Sloop of war Tarter of 26 guns on the coast of Ireland, this ship was sent out expressly to take the "Rebel Frigate Boston" commanded by Capt. Tucker (the Tarter carried six guns & thirty men <u>more</u> than the Mifflin) amongst them was twenty or thirty Volunteers, the Tarter had 12 killed, 13 wounded, the Tarter arrived in Boston & brought in with her a fine deep laden Brig prize that they captured bound for England, on our return cruise we captured two other English Brigs & recaptured a large French Ship, the ship & one Brig arrived in Boston, the other was lost at sea, the Mifflin arrived in Boston in December 1779, dismantled & hauled up for the winter.

In the spring 1780 the ship Mifflin was repaired and again fitted for sea, sailed & made a short successful cruise. I was unwell & could not go the cruise, on her return she was immediately fitted for sea again under the same officers. I then entered and was steward of the ship drawing 2-1/2 shares. We sailed from Boston the latter part of August 1780, proceeded to the Bank of Newfoundland, in Sept. captured a Letter of Marque ship carrying 18 guns. From the West Indies this ship was maned and ordered to Boston, were she arrived Oct. 1780. Immediately after dispatching, the Sept. hale commenced & lasted about eight days, after the gale was over found we were not far from Cape Hatteras. We there fell in with a scattered fleet of transports with supplies for Cornwallis Army under convoy of one Frigate, two Sloops of war, from this fleet we captured three ships with ordnances, stores, provisions, and clothing & was in chase of the fourth when we were discovered, persued & taken by the convoy ships. The Frigate proved to be the Raleigh (formerly American) Gallatea & Hyana Sloops of war, 22 & 24 guns ships. We were carried into Charleston South Carolina in the month of September, at Charleston we were put on a prison ship & kept till Jan. 1781 then exchanged and sent to Wilmington North Carolina in Jan. 1781. Sailed from Newburn N.C in the month of March as seaman in a vessel call'd the Sally, taken by Admiral Rodney's Squadron, after there taken St. Eustatia where we were bound, myself with others were put on board of the Monarch, a seventy four gun ship (Capt. Reynolds) after being transferred from ship to ship of Rodney's Squadron, then cruising off Martinico, a number of times & made to do ships duty, we at last was put on board of a prison Ship in the Island of St. Lucia, May 1781 & kept there until after the action between the French & English Fleets (Count De Grass & Admiral Hood) in this action of 4 hours the English were worsted & some of their ships much cut up, one in particular, the (Russell a line of battel ship) most of her men killed & wounded, the American prisoners with myself sixty in number were taken from the prison ship and put aboard the Russell & compelled to help work the ship to Antiqua, in going into English Harbour the Russell got on shore and never could get off. After the ship was lost we the Americans were marched across the Island from English Harbour to St. Johns, at the point of the bayonet, there imprisoned until a cartel arrived in December 1781 from Gloucester Massachusetts, when a part of us were released & arrived safe at Boston the same month. It would be needless to detail the hardship & insults which we experienced from our inveterate foe.

On our arrival home I found a new Brig building, she was soon launched & fitted for sea, carried 16 guns, was called the Genl. Scammel, Noah Stoddard commander. I entered this vessel as ships steward and acting as Liet. Of Marines. We sailed from Boston in March 1782 on a cruise in the Gulf of St. Lawrance to intercept or fall in with the Quebec fleet from England, but in proceeding on our cruise we discovered a number of people on a desolate Island (Seal Island) on our near approach we found a ship sunk. Her top gallant mast just appearing above the water. She proved to be an English Frigate, the Blonde of 32 guns having struck on a rock & sunk in a very little time. The ship cruised with a number of women that was on board had hardly time to save their lives, they landed on the Island. Two were drowned in the ship, they had not time to save anything but a scanty supply of provisions, they had been on the Island one week. We were released them from their situation, taken them on board of our Brig & landing them at a small harbour in Novia Scotia, supplied them provisions sufficient to last them to Halifax. We then proceeded on our cruise & captured a Brig from Barbadoes, sent her to Boston. Arrived in Gulf soon after, but by detonation in saving the crew of the sunken Frigate, we failed in our object in falling in with the fleet. Therefore returned to Boston in May 1782. Refitted & sailed in June to cruise off New York where we recaptured the Brig Lafayette off Boston, Capt. James Smith, a large lumber loaded ship off Newburyport & a pilot boat schooner belonging to Alexander Verginnia laden with flour & tobacco. They all arrived safe in Providence R.I. when in chase of another vessel which escaped into N.Y. the Scammel was land locked and driven on the Jersey Shore by three British ships od war just at sundown. The smallest stood in nearest, came to anchor , got springs on her cable & fired on us, a number of shots but night coming on, they hauled off shore & put out to sea, In the morning they returned in hopes to destroy our Brig, in their absence & at low water we got out our provisions with four Brass cannon & small arms which enabled us to defend our vessel & prevent their burning of her with their boats, when they found that we could defend our selves they each gave us a broad side and sheered off & put into N. York. Our damage was but trifling, rigging, some cut & one man drowned & on another wounded. After they left us & at high water we got our vessel off shore, our guns & provisions on board & arrived at Newport R.I the next day. We there repaired damage, shipped more men and sailed again in August 1782 for our old station off New York. After taking a sloop laden with flour & salt we were captured Sept 1782 by a fifty gun ship & Frigate in a very heavy blow, carried into New York, put on board the Old Jersey Prison Ship where a number of our men died. Remained on board that filthy ship until the news of peace, although promised to be released for taken in the Genl. Scammel at any time for the crew of the Blonde which we took form Seal Island, but their promises were not fulfilled till December 1782. Arrived in Boston in the month of Jan. 1783.

Author's note – Oliver Johonnot's pension application was approved and he received $114.00 per year for 1 year & 6 months as a steward & 6 months as a seaman under the 1832 Pension Act.

His time aboard the *General Mifflin* was considered United States service since it was a Massachusetts State vessel

There is no mention of his death.

George Pillsbury

Honorable S.D Ingram, Secretary of the Treasury, Washington

In consequence of the notice taken by the President in his message to Congress of the Survivors of the Revolution who had served their country and were now needy, I have been encouraged to present a narration on my services in that cause. For having been in the service of my Country either by sea or land during the whole period, having by repeated misfortunes been deprived of my property; and by an accident met with at sea on my last voyage now eight years since rendered a cripple and therefore incapable of furnishing any support either for myself or Family. I am placed in a situation to need as well as to deserve the assistance of our Government;

which I hope the following narrative may induce them to yield and which will be very gratefully received.

At the time Captain Preston ordered his "Hell Hounds" to fire on the inhabitants of Boston I was among the number then assembled in King Street, at which time Christopher Mattocks was killed near me. I aided in the destruction of the Tea destroyed in 1773, and was one of the number onboard the ship – was at the destruction of the Stamps Office. When Governor Hutchinson was treated by the Patriots, as all traitors to their country ought to be. I was rank'd among the number. When the person who inform'd against Captain Homer's cargo of wine and had il caused by the Kings Officers was apprehended I assisted in punishing him very severely in consequence of my taking so active a part in this latter transaction. I was compell'd immediately and secretly to leave Boston.

Author's note – Mr. Pillsbury was describing the famous "Boston Tea Party" and the "Boston Massacre."

From thence I went to sea, after suffering the various fortune of an inexperienced seaman for a considerable period, Captain James Scott procure'd me a passage to Boston in a transport, Captain Davidson, called the Charming Nancy, which vessel was laden with stores and troops for the British Army. General Phillips and Prescott and a number of other officers came out in the same transport. We arrived at Boston about three weeks before the Battle of Bunker Hill. I remained in Boston till September before I could make my escape to the country. After Boston was evacuated by the British I returned there and in the following June in the year 1776 I enlisted in Colonel Smith Regiment for six months which regiment was raised in Massachusetts; Our regiment arrived in New York previous to the attack of Long Island by General Clinton, the regiment on the evacuation of New York, came out with the City Guard under General Putman, when we had got breast of the road which leads from Hurl Gate to Harlem Hights we were attacked by the British and Hessian Troops, the Americans in general making a very feeble resistance, the regiment however to which I was attached made good their march to Harlem Hights, where the American Troops were forming. We work'd the whole of that night in throwing up breast works. Early the next morning we were attacked by about 7000 British and Hessians, with whom we had a smart engagement, they sustained considerable loss, retreated back to New York, our regiment to which I belonged was in the engagement at Harlem Hights. From Harlem the American Army retreated to White Plains, where our army had another engagement with the British and Hessians. From the White Plains the American Army owing to their extreme weakness was compeld to retreat off York Island, from that period until the time which I agreed to serve in the army was expired, the regiment to which I belonged was continually employed on hard and laborious duty, suffering the same hardships and privations as did the rest of the American Army.

Having been regularly discharg'd from the regiment in which I enlisted, I return'd to my friends. Believing I would better serve my country on sea than land, I went to Boston and in the following May, being 1777, I shipped on board the Privateer Sloop Satisfaction as Mate, under command of John Wheelwright Esq, mounting ten- two and three pounders, twelve swevles and sixty four men and boys, officers included. After having cruised until our provisions and water where nearly exhausted we where returning home when in Latitude 36.00 North and Long 55 West we fell in with the frigate built ship Hero of 500 tons, commanded by James Tate, laden with sugar & cotton and other produce from Tobago bound to London, mounting sixteen six pounders, twenty King arms and men in proportion.

Author's note – This location would place them off of Massachusetts/Maine.

After a warm engagement of four or five glasses, yard arm and yard arm, Captain Tate hauled down his colours and struck to Satisfaction, notwithstanding the Hero's fore was much superior to the Satisfaction; two broadsides before the hero struck I had the misfortune to receive a bad wound in the ankle, which was split several inches by a six pound shot, the bad effect of this wound I often feel even at this late period of life. After the Satisfaction return to port, not liking the vessel, I quited her and the following May being 1778 I shipped on board the Brig General Arnold as Prize Master under the command of James Magee, mounting 24 carriage Guns and 118 men and boys. Sailed on our cruise the first of June 1778, took 3 prizes, two of which was bound to Halifax, the other was bound to New York laden with wine onboard of which I was put as Prize Master and arrived safe at Boston where I remained until the Arnold was refitted for her second cruise. I entered onboard her a second time as prize master in November and sailed about 3 o'clock on the afternoon of the 24th December, the Night following our departure we experienced in the Bay a very severe snow storm, which obliged Captain Magee to put into Plymouth, where we cast anchor about 12 o'clock PM, after which the wind increased from the ENE to a perfect hurricane, after cutting away the masts and using every other exertion to save the vessel. Early on the morning of the 25th, she parted her cables and drove onshore where she soon bilged, and as the tide made filled with water, the wind attended with snow and hail continued violent until about 3 o'clock P.M., when it shifted suddenly to the NW attended with very extreme cold. Soon after which our men began to perish until 70 to 80 were frozen to death, the remainder of the officers and crew who were saved from the wreck was more or less severely injured from the effect of the frost.

From Plymouth I returned to my friends. And after having recovered from the effects of my ship wreck I entered on board the Massachusetts State Brig Tyranicide as Prize Master under command of Captain Allen Hallet, in March 1779. The Tyranicide mounted 14 four pounders and carried ninety seven men and boys. She left port the last of March cruising orders from the Board of War. Not long after sailing we fell in the Privateer Revenge, Captain Fennel, from Jamacia, mounting 16 six pounders, and from 90 to 100 men. We engaged her, after fighting her four or five glasses, with our bow sprit lashed to theirs, they struck their colors. She was first boarded by Lieutenant Cathcart & myself. We found the vessels officers and crew in a deplorable situation. Captain Fennel laying at the cabin entrance severely wounded and the first lieutenants head severed from his body. There total loss in killed and wounded 40. After capturing the Revenge we took a vessel from St. Kits laden with produce, which I was ordered to take charge of and proceed to Boston, being within 3 or 4 miles of Cape Cod, was taken by the British Sloop Ardent (Tender to the Reknown) and was carried into Newport, then put on board the Prison Ship St. Anthony. Soon after which I was relieved by a Cartel. From Providence on my return to Boston finding Capt. Hallet of the Tyranicide had been appointed to command of the State Brig Active, mounting 16 six pounders and 110 men. I entered onboard as Lieutenant and agreeable to instructions proceeded to Townsend in Maine in company with Brig Pallas, Capt. Brown and another Brig belonging to the United States to take under convoy the Transports collecting there with troops for the attack of Bragadee. After seeing the safe arrival of the Transports at Penobscot, we were ordered out by Commodore Salliston together with the other Brigs to reconnoiter the Bay. After a short time we fell in with a British fleet under command of Sir George Colliar off Owls Head, when convinced the Fleet were English, we reported to Commodore Salliston, who afterwards made a signal for every one to shift for himself, in obedience to which we proceeded up Penobscot River to Bangor, where we arrived at the head of the River in company with a number of others. The vessels were all destroy'd in obedience as ordered. The Second

Lieutenant and myself blew up the Active, after which I proceeded in company with Capt. Hallet through the woods to Boston. After travelling 5 days and suffering much for want of sustenance, we reached Fort Halifax at the head of the Kennebeck River, from thence after recovering, we reached Boston. Shortly after Captain Hallet took charge of the Brig Phenix, arme'd with 6 two pinders and eighteen men and boy's for a voyage to Grenanda. I went out his first officer. On our homeward passage off New York we took the English ship Lucy, mounting 12 guns, onboard of which I was put Prize Master. The Lucy was laden with wine and Fruit. Arrived in Boston just before the dark day, where I again met with Captain Cathcart who was fitting the Ship Essex for a cruise in the English Channel.

I shipped on board her as Third Lieutenant. Off Tory Island in Lat 55 North we took the Brig Betsey bound to Scotland.

Author's note – This would place them off of Northern Ireland.

The weather at the same time (last of November) was very boisterous and we lost two boats in boarding her. She was put under my charge as Prize Master, soon after taking command of her the Gale increased to great heights and night coming on I lost sight of the Essex. Discovering the Betsy was short on provisions I cruised for the most of the next day without finding her, then shaped my course for America. After examining the provisions and calculating that with prudence and a common chance I should reach there by the time they were expanded. The contrary was the case, owing to adverse winds our passage was very long. I at length reache'd Cape Ann Harbour with my Prize, though in a most miserable condition, having lost eleven of my men, who perished with hunger and cold.

After delivering up my prize, I entered again aboard the Essex as Third Lieutenant for a second cruise, and was taken in St. Georges Channel and carried into King Jail in Ireland. There committed to prison from which by bribing the centinal, five of us shortly after escaped, and was then retaken by a party of horse and again carried back to prison and put into the black hole for 45 days. Shortly after our release from thence, I made my escape from the prison, had to remain in Ireland 8 or 10 months and finally procured an opportunity for home via Norway and Sweden in the Robin Hood, Captain Smith, arriving at Plymouth in January 1782. When at Boston when I arrived I found Captain Cathcart commander of the Massachusetts State Ship Tartar, then fitting for sea. I agreed with Captain Cathcart to enter onboard his vessel and shortly after, 12 May 1782, I received a Commission from Governor Hancock as Second Lieutenant of the State Ship Tartar, mounting 20 nine pounders and carrying 200 men and boys, which commission I still hold. After making one cruise in which we took 3 prizes, 1 Brig, laden with salt, Brig Star, Captain Davis, of 14 guns, Brig Nonesuch, Capt. Farnam, laden with dry goods. On our return in the Tartar, the state finding her to poor to support, so expensive a vessel as the Tartar, sold her to Mess Jones and Godman, who fitted her as a Letter of Marque with 12 nine pounders and 48 men. The state having at that time no employment, I obtained leave of Governor Hancock and shipped again on board the Tartar, under Captain Cathcart as First Officer. On our passage to Virginia where we were bound for a cargo, we fell in the English Sloop of War Belesarious, Capt. Graves, of 28 guns and 200 men, being unable to get away from her, we engaged her at close quarters for 4 glasses but having several of our small number killed we were compeld to strike our colors to superior force and superior numbers. We were carried into New York and put on board the Jersey Prison Ship from whence by the influence of friends, I got parolled off Long Island, where I remained until liberated by England acknowledging the Independence of our Country.

Author's note – George Pillsbury's pension application was approved and he received $96.00 per year for 8 months and 29 days of service under the 1832 Pension Act.

His time aboard the *Tyranicide* was considered United States service since it was a Massachusetts State vessel

There is no mention of his death.

References

1. www.history.navy.mil
2. www.wikepedia.org
3. www.oldnorth.com

The National Archives Revolutionary War pension applications of;

Cornelius Arey

Joshua Atkins

Jonathan Averel

Moses Ayer

Cyprian Barnard

Benjamin Berry

Isaac Billington

William Blunt

Joseph Breed

John Brinckly

Daniel Brown

Othniel Brown

Samuel Buffum

Charles Bulkeley

Gurdin Burnham

Edmund Burr

Coggeshall Butts

Benjamin Camp

James Cassell

Giles Chester

Elkaha Cook

Enoch Crowell

Benjamin Crowninshield

Samuel Curtis

John Cushing

Luther Dana

Isaac Dewees

William Dishman

Thomas Doliber

Thomas Edgar

John Fields

Joseph Frederick

John Frisk

John Frost

Elisha Fuller

Timothy Gleeson

Reuben Godfrey

Aaron Goodwin

Richard Grinnell

John Hall

Daniel Harper

David Hawkins

James Hays

Joseph Herrington

Cyprion Henry

Daniel Hilyard

Oliver Holden

Robert Hunter

Edward Jarvis

Samuel Johnson

Oliver Johonnot

James Josiah

John Kilby

James Knight

Richard Law

William Leeds

George Long

Cheney Look

Nehemiah Manson

Henry Malcolm

Peter Masuere

Silvanus Mather

Jabez Maynard

James Mc Kinzey

John McPherson

John Manley

William Mitchel

Pierce Murphy

Henry Norris

Stephen Northup

Obed Norton

James Palmer

Asa Pease

John Peck

David Phipps

George Pillsbury

Peter Powers

Caleb Prouty

George Raymond

Nathaniel Richards

James Richardson

Vail Richmond

John Shober

Henry Skinner

Jacob Smith

Ivory Snow

Samuel Spencer

Isaac Squire

William Stacey

Moses Stanly

Mark Staples

Elijah Stevens

John Trevett

Peter Wakefield

Noah Walrond

Aaron Warren

Cornelius Wells

Jonathan Wilkins

Joseph Wilkinson

Danial Wise

John Wheeler

Esek Whipple

David White

Joseph Whitecar

Bordon Wilcox

About the Author

Ed Semler retired from the United States Coast Guard in December of 2007 with over 25 years of military service in both the United States Army and United States Coast Guard. In the United States Army he was an enlisted man and was honorably discharged as a Specialist Four (E-4). While in the United States Coast Guard he was enlisted, obtaining the rank of Master Chief Petty Officer (E-9), was commissioned as an officer, and retired as a Lieutenant (O-3E).

Fully retired he resides in Schulenburg, Texas with his wife Jana, a retired Air Force senior master sergeant. Please feel free to check out Ed's other books at www.edsemler.com, email him at mkcm378@gmail.com and check out his YouTube channel www.youtube.com/@MKCMLT

During Ed's military career he had the opportunity to sail on the Coast Guard Cutter Eagle, which is a three masted Barque. It was built by Germany in 1936 and originally named the Horst Wessel. Taken as a war prize at the end of World War II it is now used to train future Coast Guard officers and is homeported at New London, Connecticut. Ed sailed on her during her 3-month summer cruise in 1998 from San Juan, Puerto Rico, up the east coast to Halifax, Nova Scotia, and back to New London.

USCGC Eagle on her summer 1998 cruise

His other publications are;

"Around The World," a memoir of his 25 years of service as an officer and enlisted man in the U.S. Army and U.S. Coast Guard

"U.S. Coast Guard Cutter Sherman (WHEC-720) Circumnavigation Deployment 2001" which details the *Sherman's* historic circumnavigation of the globe and deployment to the Persian Gulf in 2001

"The Three Gunsallus Brothers" a story about fighting for Pennsylvania during the Civil War

"Sam Houston & Napoleon Bonaparte Meet On The Civil War Battlefield" a true story of the Walker brothers

"Thoughts On Being A Chief Petty Officer" a take on military leadership

"Fighting For Pennsylvania In The Early Years 1763 to 1783 – The Story of Captain Thomas Askey And Lieutenant Richard Gunsalus Of Cumberland County"

"Joe Semler Playing Baseball in the 1920's &30's"

"Alice Springs Australia Adventures In The 80's"

"Count On Us Coast Guard Cutter Dependable – Law Enforcement And Search & Rescue"

"United States Coast Guard Tragedies"

“In Their Own Words – Short Stories of Pennsylvanians in the Revolutionary War”

www.ingramcontent.com/pod-product-compliance
Lightning Source LLC
LaVergne TN
LVHW051541080426
835510LV00020B/2809

* 9 7 8 1 7 3 7 6 4 7 2 3 2 *